Lecture Notes in Mathematics

A collection of informal reports and seminars
Edited by A. Dold, Heidelberg and B. Eckmann, Zürich

45

Albert Wilansky

Lehigh University, Bethlehem, Pennsylvania

Topics in Functional Analysis

Notes by W. D. Laverell

1967

Springer-Verlag · Berlin · Heidelberg · New York

Library of Congress Catalog Card Number 67-30514. Printed in Germany. Title No. 7365

Introduction

Several topics are touched on in these notes.

1. When spaces are continuously embedded in a fixed Hausdorff space, the identity map between them has a closed graph. Since this allows use of forms of the closed graph theorem, one tries to see when a collection of given spaces can be so embedded. Since they are automatically continuously embedded in their inductive limit, the question of when the inductive limit is separated arises. Since the inductive limit is a quotient of the direct sum, an obvious criterion for separation is at hand. (See Chapter 2, Section 8). For two spaces, this criterion reduces to the familiar one that the identity map between them have closed graph. Thus we obtain the insight that the graph of the identity map should be considered as closed in the direct sum, rather than in the product of two spaces, when it is desired to generalize this concept to more than two spaces.

2. To round out the discussion of embedded spaces and closed graph, we have included some material on comparison of topologies (Chapter 1) including a new type of space in which sequential convergence is trivial (Section 4). Also lattice properties and completions of embedded spaces are discussed.

3. For wider applicability of the results, local convexity was not always assumed. This leads to what we call the unrestricted inductive limit which is discussed in Chapter 2. Removing local convexity from the concept of barreled, we have

some ideas introduced by the author and some by Wendy Robertson in Chapter 3. Also shown in Chapter 3 are some connections between bornological and sequential spaces. The digression into sequential spaces was suggested by the trivial sequential convergence shown in Chapter 1, Section 4.

4. Finally an important result of my former student, A. K. Snyder, on conull FK spaces is presented in Chapter 6. For use in proving Snyder's theorem we present a self-contained version of parts of the two-norm theory recently invented by the Polish school.

Statement of Novelty of Results

Much of the material in these notes is expository. The author claims as new only Theorem 1.4, p. 2; Chapter 1, Section 2; Lemma 3.2, p. 12; Chapter 1, Section 4, except for the first two lines; Chapter 2, Section 3; Chapter 2, Sections 7,8; Example 1.1, p. 45; Theorem 1.1, Example 1.3, and Corollary, pp. 46-48; Examples 2.1, 2.2, pp. 52-53; Theorem 2.4, p. 54; Theorem 2.5, p. 55; the proof on p. 56; pp. 57-58; Example 2.8, p. 60; Example 1.1, p. 63; pp. 65-66; Example 3.5, p. 72; Chapter 5, Section 2; Example 3.1, p. 80; Example 3.2, p. 81.

Acknowledgements

Example 2.5, p. 8 is due to my student P. G. Jessup. W. H. Ruckle made many valuable suggestions all of which have been incorporated into these notes. I wish to thank my student W. D. Laverell for writing the notes and Rosemarie Ehser for her dedication and skill in typing them.

Chapter One

Comparison of Topologies

§1 Maps with Closed Graph

Let x and Y be topological spaces, $D \subset X$, and let $f:D \to Y$. We will say f is a map from X to Y. We say that f from X to Y has closed graph if the graph of f is closed in $X \times Y$.

For example, let $X = (1,3)$, $Y = (2,4)$, and let i from X to Y be the identity, that is $i: (2,3) \to Y$. Then i from X to Y has closed graph, but the graph of i is not closed in $R \times R$.

Example 1.2. Define u from C[0,1] to C[0,1] by $u(f) = f'$, where C[0,1] has the norm given by $\|f\| = \sup\left\{|f(x)| : x \in [0,1]\right\}$. Then u has closed graph.

Since a pair (x,y) is a member of the graph of a map f from X to Y if and only if $y = f(x)$, and since a net (x_δ, y_δ) n $X \times Y$ converges to a point $(x,y) \in X \times Y$ if and only if $x_\delta \to x$ in X and $y_\delta \to y$ in Y, we have the following characterization of the closed graph property:

Theorem 1.1. <u>A map f from X to Y has closed graph if and only if for any net x_δ in X converging to a point $x \in X$ with $f(x_\delta) \to y \in Y$, $y = f(x)$.</u>

Theorem 1.2. <u>Let X and Y be overlapping topological spaces ($X \cap Y \neq 0$). Let f from X to Y be the identity and let g from Y to X be the identity. The following are pairwise equivalent:</u>

i. *f has closed graph*

ii. *g has closed graph*

iii. *if* $x \in X$, $y \in X$, *and* $x \neq y$, *there exist neighborhoods* $U \subseteq X$ *of* x *and* $V \subseteq Y$ *of* y *such that* $U \cap V = \phi$

i $\Longleftrightarrow$ ii. symmetry

i $\longrightarrow$ iii. Let f have closed graph, G. Let x and y be as given in iii. Then $(x,y) \notin G$. Therefore, there exists a basic neighborhood W of (x,y) not meeting G. Then $P_X[W]$ and $P_Y[W]$ are the required neighborhoods.

iii $\longrightarrow$ i. Let $(x,y) \notin G$. Choose U, V as in iii and let $W = U \times V$. Then W is a product neighborhood of (x,y) not meeting G.

Theorem 1.3. *Let* X *and* Y *be overlapping spaces. If* $X \cup Y$ *has a Hausdorff topology smaller than the initial topologies of* X *and* Y, *then* i, *the identity from* X *to* Y, *has closed graph.*

Proof: Let $x_\delta \to x \in X$ with $x_\delta \in X \cap Y$. Let $x_\delta \to y \in Y$. Then $x_\delta \to x$ and $x_\delta \to y$ in the smaller Hausdorff topology. Hence, $y = x$.

The following example shows that it is not sufficient that $X \cap Y$ have a smaller Hausdorff topology for i to have closed graph.

Example: $X = [0,1]$, $Y = [0,1) \cup \{2\}$. X has the ordinary topology, and Y has the ordinary topology on $[0,1)$ while neighborhoods of 2 are of the form $(x,1) \cup \{2\}$.

In the setting of a linear topological space we have the following converse of Theorem 1.3.

Theorem 1.4. *Let* X *and* Y *be linear topological spaces which are linear subspaces of a linear space* L. *Suppose* i *from* X *to* Y

has closed graph. Then $X + Y$ *has a linear separated topology smaller on* X *and on* Y *than the initial topologies.*

Proof: Let $\mathcal{U} = \{U + V : U$ is a balanced open neighborhood of 0 in X, V is a balanced open neighborhood of 0 in $Y\}$. $\mathcal{U}$ generates a linear topology for $X + Y$ ([13], §10.1, Theorem 3). This topology is separated, for let $z \neq 0$, say $z = x-y$, $x \neq y$. By Theorem 1.2, there exists an open neighborhood U of 0 in X, and an open neighborhood V of 0 in Y such that $(x - U) \cap (y + V) = \emptyset$. Therefore, $z = x - y \notin U + V \in \mathcal{U}$, for if $x - y = x' + y'$, $x' \in U$, $y' \in V$, then $x - x' = y + y'$, and $(x - U) \cap (y + V) \neq \emptyset$.

The $\mathcal{U}$ topology is smaller on X than the initial topology of X, for let $U + V \in \mathcal{U}$. $(U + V) \cap X$ is open in X since it is a union of translates of an open set, U, in X.

At this point we remark that for a linear map f from X to Y, the condition that $x_\delta \to 0$, $f(x_\delta) \to y \in Y \implies y = 0$ is not sufficient for f to have closed graph since f may not be defined on all of X.

Theorem 1.5. (V. Ptak, [16], Theorem 3.7)

Let X *and* Y *be locally convex linear topological spaces, and let* $T : X \to Y$ *be a linear map. Then* T *has closed graph if and only if* $T^{*-1}[X']$ *is total over* Y.

Proof: Necessity - Let $z \in Y$, $z \neq 0$. There exists U, a convex, balanced neighborhood of 0 in X with $z \notin \overline{T[U]}$. [Choose a basic neighborhood of $(0,z) \in X \times Y$ not meeting the graph of T, and let U be its projection into X.] Therefore, there exists

$g \in Y'$ with $|g(y)| \leq 1$ for all $y \in T[U]$, and with $|g(z)| > 1$. $T^*g = g \circ T \in X'$ since it is bounded on U.

Sufficiency - Let $x_n \to 0$ in X, $T(x_n) \to y \in Y$. Let $g \in T^{*-1}[X']$. $g(y) = \lim_{n\to\infty} g(T(x_n)) = \lim_{n\to\infty} T^*g(x_n) = 0$. Thus, $y = 0$.

§2. Linear Inf and Intersection of Two Topologies.

In this section we make the blanket assumption that $\mathcal{J}$ and $\mathcal{J}'$ are topologies for a set X.

Theorem 2.1. If $\mathcal{J}$ and $\mathcal{J}'$ are T_1, so is $\mathcal{J} \cap \mathcal{J}'$.

Proof: $x \in X$ implies that $\{x\}^\sim \in \mathcal{J} \cap \mathcal{J}'$.

Example 2.1. A $\mathcal{J}$ and a $\mathcal{J}'$ neighborhood of x which is not a $\mathcal{J} \cap \mathcal{J}'$ neighborhood of x.

Let $X = R^2$, $p(x,y) = |x|$, $q(x,y) = |y|$, and let $\mathcal{J}$ and $\mathcal{J}'$ be the associated topologies. Then $\mathcal{J} \cap \mathcal{J}'$ is indiscrete. Take $U = (p<1) \cup (q<1)$.

Example 2.2. A $\mathcal{J}$ and a $\mathcal{J}'$ neighborhood of x which is not a $\mathcal{J} \cap \mathcal{J}'$ neighborhood of x where $\mathcal{J}$ and $\mathcal{J}'$ are normed topologies.

Let $\mathcal{J}$ and $\mathcal{J}'$ be the topologies associated with two non-comparable norms, p and q. Let V be a $\mathcal{J} \cap \mathcal{J}'$ open neighborhood of 0. Then V includes $(p<e)$ for some $e > 0$. q is not bounded on $(p < e)$ for if so p is stronger than q ([13], §4.2, Fact vi). Choose $x \in (p<e)$ with $q(x) > 2$. Then $x \in V$ so V is a $\mathcal{J}'$ neighborhood of x. Hence, V includes a $\mathcal{J}'$ sphere of radius δ, $0 < \delta < 1$, centered at x. p is not bounded on this sphere. Choose y in it with $p(y) > 1$.

Now $q(y) \geq q(x) - q(x-y) > 1$. Let $U = (p < 1) \cup (q < 1)$. U is a $\mathcal{T}$ and $\mathcal{T}'$ neighborhood of 0, but is not a $\mathcal{T} \cap \mathcal{T}'$ neighborhood of 0.

Example 2.3. $\mathcal{T}$ and $\mathcal{T}'$ may be T_2, T_3, T_4 topologies for countable, compact subsets of R and yet $\mathcal{T} \cap \mathcal{T}'$ may not be T_2.

Let $\mathcal{T}_n$ be the no point compactification of N (the integers) at n (i.e. the one point compactification of $N - \{n\}$ with $n = \infty$). $\mathcal{T}_1 \cap \mathcal{T}_2$ is not T_2, for let U and V be $\mathcal{T}_1 \cap \mathcal{T}_2$ neighborhoods of 1 and 2 respectively. Since U is a $\mathcal{T}_1$ neighborhood of 1, it contains all but finitely many integers, and similarly so does V. Hence, $U \cap V \neq \emptyset$.

Example 2.4. $\mathcal{T}$ and $\mathcal{T}'$ linear topologies for a linear space. $\mathcal{T} \cap \mathcal{T}'$ not a linear topology.

Let p and q be non-comparable complete norms for a linear space X, and let $\mathcal{T}$ and $\mathcal{T}'$ be the associated topologies. I assert that $\mathcal{T} \cap \mathcal{T}'$ is not T_2. If it were, $i: (X, \mathcal{T}) \to (X, \mathcal{T}')$ would have closed graph by Theorem 1.3, but then i would be continuous by the Closed Graph Theorem.

By Theorem 2.1, $\mathcal{T} \cap \mathcal{T}'$ is T_1. Hence, it cannot be a linear topology.

Theorem 2.2. Let $\mathcal{T}$ and $\mathcal{T}'$ be linear topologies for a linear space X. In $\mathcal{T} \cap \mathcal{T}'$ addition and multiplication are separately continuous.

Proof: Addition - Fix $y \in X$ and let $x_\delta \to x$. Let U be a $\mathcal{J} \cap \mathcal{J}'$ open neighborhood of $x + y$. Then $U - y$ is a $\mathcal{J} \cap \mathcal{J}'$ neighborhood of x. Hence, $x_\delta \in U - y$ eventually, and $x_\delta + y \in U$ eventually.

Multiplication - Fix a vector $x \in X$. Let $t_\delta \to t$ in K. Then $t_\delta x \to tx$ in $\mathcal{J}$, therefore in $\mathcal{J} \cap \mathcal{J}'$.

Next fix a non-zero scalar t, and let $x_\delta \to 0$ in $\mathcal{J} \cap \mathcal{J}'$. Let U be a $\mathcal{J} \cap \mathcal{J}'$ open neighborhood of 0. Then $\frac{1}{t} U$ is a $\mathcal{J} \cap \mathcal{J}'$ neighborhood of 0. Therefore $x_\delta \in \frac{1}{t} U$ eventually, and so $tx_\delta \in U$ eventually.

Finally, let $x_\delta \to x$. Then $tx_\delta = t(x_\delta - x) + tx \to tx$ [$(x_\delta - x) \to 0$ by the first part of the proof].

In Example 2.4, addition in $\mathcal{J} \cap \mathcal{J}'$ is not continuous, for if it were, $(X, \mathcal{J} \cap \mathcal{J}', +)$ would be a topological group, hence would be T_2 since it is T_1.

Theorem 2.3. Let $\mathcal{J}$ and $\mathcal{J}'$ be linear topologies for a linear space X. Then $\mathcal{J} \cap \mathcal{J}'$ is T_2 if and only if $i: (X, \mathcal{J}) \to (X, \mathcal{J}')$ has closed graph.

Proof: Theorems 1.3 and 1.4.

If in this theorem we omit the hypothesis that $\mathcal{J}$ and $\mathcal{J}'$ be linear topologies, we still have the result that the condition "i has closed graph" is necessary. We may also obtain the following:

Theorem 2.4. Let $\mathcal{J}$ and $\mathcal{J}'$ be T_2 topologies for X with $i: (X, \mathcal{J}) \to (X, \mathcal{J}')$ having closed graph. For $x \neq y$, there exists a $\mathcal{J}$ and a $\mathcal{J}'$ neighborhood, U, of x, and a $\mathcal{J}$ and a $\mathcal{J}'$ neighborhood, V, of y such that $U \cap V = \emptyset$.

Proof: Let U and V' be respectively a $\mathcal{T}$ neighborhood of x and a $\mathcal{T}'$ neighborhood of y with $U \cap V' = \emptyset$. Let U' and V be respectively a $\mathcal{T}'$ neighborhood of x and a $\mathcal{T}$ neighborhood of y with $U' \cap V = \emptyset$. Let W and N be disjoint $\mathcal{T}$ neighborhoods of x and y respectively and W' and N' disjoint $\mathcal{T}'$ neighborhoods of x and y respectively.

Let $A = U \cap W$, $A' = U' \cap W'$, $B = V \cap N$, and $B' = V' \cap N'$. Then $A \cup A'$ and $B \cup B'$ are the required neighborhoods. That they are disjoint follows from the equation

$$\begin{aligned}(A \cup A') \cap (B \cup B') &= (A \cap B) \cup (A \cap B') \cup (A' \cap B) \cup (A' \cap B') \\ &\subset (W \cap N) \cup (U \cap V') \cup (U' \cap V) \cap (W' \cap N') = \emptyset.\end{aligned}$$

Lemma 2.1. *Let* $\mathcal{T}$ *and* $\mathcal{T}'$ *be linear topologies for a linear space. Let* U *be a* $\mathcal{T}$ *open neighborhood of* 0 *and* V *a* $\mathcal{T}'$ *open neighborhood of* 0. *Then* $U + V$ *is a* $\mathcal{T} \cap \mathcal{T}'$ *open neighborhood of* 0.

Proof: $U + V = \bigcup \{u + V\colon u \in U\} = \bigcup \{U + v\colon v \in V\}$, hence is a $\mathcal{T}$ and a $\mathcal{T}'$ open neighborhood of 0.

Theorem 2.5. *Let* $\mathcal{T}$ *and* $\mathcal{T}'$ *be linear topologies for a linear space* X. *Then* $\mathcal{T} \cap \mathcal{T}'$ *is linear if and only if for every* $\mathcal{T} \cap \mathcal{T}'$ *neighborhood* U *of* 0 *there exists a* $\mathcal{T}$ *neighborhood* V *and a* $\mathcal{T}'$ *neighborhood* W *of* 0 *with* $V + W \subset U$.

Proof: Necessity - Trival.

Sufficiency - We check the conditions of [13], §10.1, Theorem 1.

(a) Trivial

(b) Trival

(c) Let U be a $\mathcal{J} \cap \mathcal{J}'$ neighborhood of 0. Let V and W be balanced $\mathcal{J}$ and $\mathcal{J}'$ open neighborhoods of 0 respectively with $V + W \subset U$. By Lemma 2.1, $V + W$ is a $\mathcal{J} \cap \mathcal{J}'$ neighborhood of 0.

(d) Choose V and W as in the hypothesis. Let V_1 and W_1 be $\mathcal{J}$ and $\mathcal{J}'$ open neighborhoods of 0 with $V_1 + V_1 \subset V$, $W_1 + W_1 \subset W$. Then $V_1 + W_1$ is a $\mathcal{J} \cap \mathcal{J}'$ neighborhood of 0 with $V_1 + W_1 + V_1 + W_1 \subset V + W \subset U$.

Example 2.5. Let $(X, \mathcal{J})$ be a linear topological space, and let $\mathcal{J}' = \sigma(X, X^\#)$. Then $\mathcal{J} \wedge \mathcal{J}' = \sigma(X, X')$.

Proof: A typical neighborhood of 0 in $\mathcal{J} \wedge \mathcal{J}'$ is $U + V$, where $U = \bigcap_{i=1}^{n} (|f_i| < \epsilon)$ and V is a $\mathcal{J}$ neighborhood of 0. $U \supset \bigcap_{i=1}^{n} f_i^{\perp} = L$. Let W be a $\mathcal{J}$ neighborhood of 0 with $W + W \subset V$. Then $V + L \supset W + W + L \supset W + \bar{L}$. Let $X = \bar{L} + a_1 + \ldots + a_m$. Every $x \in X$ may be uniquely represented as $x = \sum_{i=1}^{m} \alpha_i a_i + \ell$, with $\ell \in \bar{L}$. Define $g_i(x) = \alpha_i$. $g_i^{\perp} \supset \bar{L}$, so g_i is the sum of $\bar{L}$ and some finite dimensional subspace. Hence, $g_i^{\perp}$ is closed, and so g_i is continuous.

Let N be a $\mathcal{J}$ neighborhood of 0 with $\sum_{i=1}^{m} N \subset W$. Choose $\epsilon > 0$ such that $|t| < \epsilon$ implies $ta_i \in N$ for all i. Finally, $\bigcap_{i=1}^{m} (|g_i| < \epsilon) \subset U + V$, because if $g_i(x) < \epsilon$ for each i we have $x = \sum_{i=1}^{m} \alpha_i a_i + \ell \in \sum_{i=1}^{m} N + \bar{L} \subset W + \bar{L} \subset V + L \subset U + V$.

The converse inequality is trivial.

Example 2.6. Let X be a linear space and $\mathcal{T}$ a linear separated topology for X with $(X,\mathcal{T})' = \{0\}$. Then $\sigma(X^{\#}) \wedge \mathcal{T}$ is indiscrete, while $\sigma(X^{\#}) \cap \mathcal{T}$ is T_1.

Theorem 2.6. *Let* $\mathcal{T}$ *and* $\mathcal{T}'$ *be locally convex linear topologies for a linear space* X. *Then* $\mathcal{T} \cap \mathcal{T}'$ *is locally convex if and only if it is a linear topology.*

Proof: Necessity - Let U be a $\mathcal{T} \cap \mathcal{T}'$ neighborhood of 0. By hypothesis, there exists a convex $\mathcal{T} \cap \mathcal{T}'$ neighborhood V of 0 such that $V \subset U$. V is a $\mathcal{T}$ neighborhood of 0, so $\frac{1}{2} V$ is also. Similarly, $\frac{1}{2}V$ is a $\mathcal{T}'$ neighborhood of 0. $\frac{1}{2} V + \frac{1}{2} V = V$.

Sufficiency - Let U be a $\mathcal{T} \cap \mathcal{T}'$ neighborhood of 0. Let V be a $\mathcal{T} \cap \mathcal{T}'$ neighborhood of 0 with $V + V \subset U$. There exists a convex $\mathcal{T}$ neighborhood W of 0 with $W \subset V$ and a convex $\mathcal{T}'$ neighborhood W' of 0 with $W' \subset V$. Then $W + W'$ is a convex $\mathcal{T} \cap \mathcal{T}'$ neighborhood of 0 (Lemma 2.1), and $W + W' \subset V + V \subset U$.

Theorem 2.7. *If* $\mathcal{T}$ *and* $\mathcal{T}'$ *are locally convex linear topologies,* $\mathcal{T} \wedge \mathcal{T}'$ *is locally convex.*

Proof: Same proof as sufficiency in Theorem 2.6.

Remark: If $\mathcal{T}$ and $\mathcal{T}'$ are linear topologies and $\mathcal{T} \cap \mathcal{T}'$ is not linear, there exists a $\mathcal{T} \cap \mathcal{T}'$ neighborhood U of 0 such that $0 \in \cap\{\tilde{U} + V + W\colon V$ is a $\mathcal{T}$ neighborhood of 0, W is a $\mathcal{T}'$ neighborhood of $0\} = \cap\{cl_{\mathcal{T}}(\tilde{U} + W)\colon W$ is a $\mathcal{T}'$ neighborhood of $0\}$.

Proof: Choose U so that $V + W \not\subset U$ for every $\mathcal{T}$ neighborhood V of 0 and every $\mathcal{T}'$ neighborhood W of 0 (by Theorem 2.5), that is $\tilde{U}$ meets $V + W$. Then $0 \in \tilde{U} - V - W$.

§3 Almost Open Maps

Definition: *Let X and Y be topological spaces. A map f from X to Y is called almost open if for every point $x \in X$ and every neighborhood U of x, $\overline{f[U]}$ is a neighborhood of f(x).*

Lemma 3.1. *Let X and Y be topological groups and f a homomorphism from X to Y. Suppose for every open set $G \subset X$, f[G] is somewhere dense. Then f is almost open.*

Proof: Let U be a neighborhood of 0. Let V be an open neighborhood of 0 with $V-V \subset U$. Then $\overline{f[U]} \supset \overline{f[V] - f[V]} \supset \overline{f[V]} - \overline{f[V]}$. Hence $\overline{f[U]}$ is a neighborhood of 0.

Example 3.1. Let $(X,\mathcal{J})$ be a linear topological space of category II. Then every linear map of a linear topological space onto X is almost open.

Proof: Let U be a neighborhood of 0 in the domain of f. Then $\cup \{n\, f[U]\} = X$. Therefore f[U] is somewhere dense.

Example 3.2. Let $\mathcal{J}$ and $\mathcal{J}'$ be locally convex linear topologies with the same closed convex sets (e.g. with the same dual). If $i: (X,\mathcal{J}) \to (X,\mathcal{J}')$ is almost open, then it is open. That is, $\mathcal{J}'$ is stronger than $\mathcal{J}$.

The preceding results give a standard form of the open mapping theorem.

Definition: *Let X and Y be topological spaces and f a map from X to Y. Call f thickening if for every open set $U \subset X$, f[U] is everywhere dense.*

It is clear that thickening implies almost open. A reference for the following discussion may be found in [6], page 240.

Example 3.3. Let X be a linear space, $\mathfrak{J}$ a linear topology for X with $(X,\mathfrak{J})' = \{0\}$. Let $S \subset X^{\#}$. Then $i: (X,\mathfrak{J}) \to (X,\sigma(S))$ and its inverse are both thickening: this means that for every $\mathfrak{J}$ open set U and every $\sigma(S)$ open set V, $U \cap V \neq \emptyset$.

Proof: Suppose $U \cap V = \emptyset$, where U is $\mathfrak{J}$ open and V is $\sigma(S)$ open. Then V contains a linear subspace of finite codimension. Let M be the $\mathfrak{J}$ closure of that linear subspace. Then $M \cap U = \emptyset$. Let L be a maximal linear subspace of X which includes M. Then L is $\mathfrak{J}$ closed. Hence $L = f^{\perp}$ with $f \in (X,\mathfrak{J})'$.

Example 3.4. Suppose $(X,\mathfrak{J})' = \{0\}$ and that $(X,\mathfrak{J})$ is of category II. Let $\mathfrak{J}'$ be a locally convex topology for X. Then $i: (X,\mathfrak{J}) \to (X,\mathfrak{J}')$ and its inverse are both thickening.

Proof: Let U be a convex $\mathfrak{J}'$ neighborhood of 0. Then $\bigcup_{n=1}^{\infty} \{n\ U\} = X$. Thus, U is somewhere dense with respect to $\mathfrak{J}$. Therefore, $cl_{\mathfrak{J}}\ U$ is a convex $\mathfrak{J}$ neighborhood of 0. Therefore, $cl_{\mathfrak{J}}\ U = X$.

Example 3.5. Let X be a topological group which is either separable or Lindelof. Let Y be a topological group of category II, and let f be a homomorphism of X onto Y. Then f is almost open.

Proof: Let U be a neighborhood of 0 in X. Then there exists a countable set C with $C + U = X$. Then $f[C] + f[U] = Y$, that is $Y = \bigcup \{t + f[U] : t \in C'\}$ with C' countable. Thus, f[U] is somewhere dense.

Example 3.6. Let X be a locally compact group, Y a separated group, and f a continuous homomorphism from X onto Y. Then if f is almost open, f is open.

Proof: Let U be a compact neighborhood of 0 in X. Then f[U] is compact, hence closed. Therefore, $f[U] = \overline{f[U]}$ has interior.

Application: Two comparable locally compact separated separable group topologies are equal.

Proof: i is a continuous homomorphism, hence is almost open by example 3.5, thus is open by example 3.6.

Lemma 3.2 *Let X,Y and Z be topological spaces and* $f: X \to Y$, $g: Y \to Z$. $g \circ f$ *is almost open if either f is open and g is almost open or f is almost open and g is continuous and almost open.* $g \circ f$ *is not necessarily open if f and g are almost open. In our first counter-example f is a homeomorphism into, and g is open, 1-1, and onto. In our second counter-example f and g are both linear, 1-1, and onto.*

Proof: First part is trivial.

Let f be almost open, g continuous and almost open. Let U be a neighborhood of $x \in X$. Then $\overline{g \circ f[U]} = \overline{g[\overline{f[U]}]} \supset g[\overline{f[U]}]$ Therefore, $\overline{g \circ f[U]} \supset \overline{g[\overline{f[U]}]}$.

First counter example: Y = [0,1] X = rationals in [0,1], Z = Y with X adjoined as a closed set. $i: X \to Y$ and $i: Y \to Z$ are both almost open, but $i: X \to Z$ takes X into a closed set with empty interior.

Second Counter-example: Let $(X,\mathfrak{J})' = \{0\}$ with $(X,\mathfrak{J})$ of category II. Let $S \subset X^{\#}$ be such that $(X,\sigma(S))$ is not a barreled space [e.g. let p be a norm on X, and let S be $(X,p)'$].

Consider

$$\tau(X^{\#})) \xrightarrow{i} (X,\mathfrak{J}) \xrightarrow{i} (X,\sigma(S)).$$

where τ is the Mackey topology, the strongest possible locally convex topology.

i: $(X,\tau(X^{\#}) \to (X,\sigma(S))$ is not almost open, for let B be a $\sigma(S)$barrel which is not a $\sigma(S)$ neighborhood of 0. It is a τ barrel since τ is stronger. But B is nowhere dense in $(X,\sigma(S))$.

Remark: A category II linear topology is larger than any linear topology $\mathcal{V}$ having a local base of $\mathfrak{J}$ closed neighborhoods of 0.

Proof: Let U be a $\mathfrak{J}$ closed $\mathcal{V}$ neighborhood of 0. $X = \bigcup_{n=1}^{\infty} \{nU\}$, so U is $\mathfrak{J}$ somewhere dense. Since U is $\mathfrak{J}$ closed, it is a $\mathfrak{J}$ neighborhood of 0.

Definition: <u>A linear topological space</u> $(X,\mathfrak{J})$ <u>is called ultra-barreled if every linear topology</u> $\mathcal{V}$ <u>for</u> X <u>having a local base of</u> $\mathfrak{J}$ <u>closed sets is smaller than</u> $\mathfrak{J}$.

See [3] , page 427.

Lemma 3.3. <u>Let</u> X <u>be a</u> T_2 <u>space and</u> $\{y_n\}$ <u>be a sequence in</u> X. <u>Suppose that</u> $\{y_n\}$ <u>is eventually in every closed neighborhood of</u> $x \in X$. <u>Then either</u> $\{y_n\}$ <u>is divergent or converges to</u> x.

The next result is due to J. D. Weston [12] , Theorem 8, p. 347.

Theorem 3.1. <u>Let</u> X <u>be a complete metric space and</u> Y <u>a</u> T_2 <u>space</u>. <u>Let</u> $f:X \to Y$ <u>be one to one, continuous, and almost open</u>. <u>Then</u> f <u>is a homeomorphism</u>.

Proof: We shall show the following: (1) $f[N(x,2\epsilon)] \supset \overline{f[N(x,\epsilon/2)]}$ for all $x \in X$, $\epsilon > 0$. We shall define sequences $\{x_n\}$ and $\{t_n\}$ in X, and we define for convenience

$$U_n = f[N(x_n, \epsilon/2^n)]$$

$$V_n = f[N(t_n, \epsilon/2^n)]$$

Let $x_1 = x$. To prove (1), let $y \in RHS$, say $y = f(t_1)$. $y \in \overline{U}_1$, so every neighborhood of y meets U_1. In particular, $\overline{V}_1$ meets U_1. Choose $x_2 \in N(x_1, \epsilon/2)$ with $f(x_2) \in \overline{V}_1$. Since $f(x_2) \in \overline{V}_1$, every neighborhood of $f(x_2)$ meets V_1. In particular, $\overline{U}_2$ meets V_1. Choose $t_2 \in N(t_1, \epsilon/2)$ with $f(t_2) \in \overline{U}_2$. Suppose we have chosen $x_1, \ldots, x_{n-1}$, $t_1, \ldots, t_{n-1}$ with $x_i \in N(x_{i-1}, \epsilon/2^{i-1})$, $f(x_i) \in \overline{V}_{i-1}$, $t_i \in N(t_{i-1}, \epsilon/2^{i-1})$, and $f(t_i) \in \overline{U}_i$ for $i = 2, \ldots, n$.

We get x_n and t_n by repeating the first step with 1 replaced by n-1, 2 replaced by n.

$d(x_i, x_{i-1}) < \epsilon/2^{i-1}$, so x_n is a Cauchy sequence. Similarly, x_n is a Cauchy sequence, Let $x_n \to \ell$. We shall see that $x_n \to \ell$. Since f is one to one and continuous, it is sufficient to show that $f(t_n) \to f(\ell)$. Since f is continuous, $\{f(t_n)\}$ is convergent. By lemma 3.3 it is sufficient to show that $\{f(t_n)\}$ is eventually in every closed neighborhood V of $f(\ell)$. $F^{-1}[V]$ is a closed neighborhood of ℓ. Therefore $N(x_n, \epsilon/2^n) \subset f^{-1}[V]$ eventually. Therefore, $U_n \subset V$ eventually, so $\overline{U}_n \subset V$ eventually. Therefore, $f(t_n) \in V$ eventually.

Now, $d(t_1, x_1) \leq d(t_1, \ell) + d(x_1 \ell_1)$

$$\leq \sum_{i=1}^{\infty} \frac{\epsilon}{2^i} + \sum_{i=1}^{\infty} \epsilon/2^i$$

$$= 2\epsilon$$

Thus, $t_1 \in N(x, 2\epsilon)$, and $y \in$ L.H.S.

Example 3.7. The inclusion map from the irrationals to the real numbers shows that we cannot drop the hypothesis that f is onto in Theorem 3.1, the irrationals being completely metrizable.

We remark that the assumption that f is one to one may be dropped in case f may be written as $h \circ g$ where $g: X \to S$ is continuous, onto, and open; $h: S \to Y$ is one to one; and S is a complete metric space.

Example 3.8. Let X be a Frechet space and Y a linear separated space. If $f: X \to Y$ is continuous and almost open, then f is open.

Proof: If f is onto we may apply the preceding remark with $S = X/f^{\perp}$.

If f is one to one, $f: X \to f[X]$ is open, hence a linear homeomorphism. Thus, $f[X]$ is complete. But $f[X]$ is dense since f is almost open, hence $f[X] = Y$.

If f is neither one to one nor onto we apply the preceeding to the induced map from $X/f^{\perp} \to Y$.

Remark: An almost open map into a T_1 space preserves isolated points.

Lemma 3.4. $f: (X,\mathcal{J}) \to (Y,\mathcal{J}')$ <u>is almost open if and only if</u> $f: (X,\mathcal{J} \vee f^{-1}(\mathcal{J}')) \to (Y,\mathcal{J}')$ <u>is almost open</u>.

Proof: Sufficiency - Trivial

Necessity - Let U be a $\mathcal{J} \vee f^{-1}(\mathcal{J}')$ neighborhood of x. Then $U \supset V \cap f^{-1}[W]$ where V is a $\mathcal{J}$ neighborhood of x and W is a $\mathcal{J}'$ open neighbrohood of $f(x)$. Then $\overline{f[U]} \supset \overline{f[V] \cap W} \supset \overline{f[V]} \cap W$.

Lemma 3.5. <u>Let</u> $\mathcal{A}$ <u>be a chain of topologies on a set</u> X. <u>Let</u> $f: (x,\mathcal{J}) \to Y$ <u>be almost open for every</u> $\mathcal{J} \in \mathcal{A}$. <u>Then</u> $f: (X, \vee \mathcal{A}) \to Y$ <u>is almost open</u>.

Proof: Let U be a $v\mathcal{A}$ neighborhood of $x \in X$. There exists $\mathcal{J} \in \mathcal{A}$ such that U is a $\mathcal{J}$ neighborhood of x. Therefore, $\overline{f[U]}$ is a neighborhood of f(x) in Y.

Theorem 3.2. Let X and Y be complete metric spaces. Let $f:X \to Y$ be one to one, onto, almost open, and have closed graph. Then f is open.

[If we assume that f is one to one, almost continuous, and has closed graph the same proof will show that f is continuous].

Proof: Let d and d' be the metrics for X and Y. Consider $f:(X,d+(d' \circ f)) \to (Y,d')$. By lemma 3.4 f is almost open. f is clearly continuous. Finally, $d + (d' \circ f)$ is a complete metric for X, for let $\{x_n\}$ be a Cauchy sequence in X. Then $\{x_n\}$ is a d Cauchy sequence, so $x_n \to x$ in (X,d). Also $f(x_n)$ is a d' Cauchy sequence, say $f(x_n) \to y$ in (Y,d'). Since f has closed graph, $y = f(x)$. Thus, $d(x_n,x) + d'(f(x_n),f(x)) \to 0$, that is $x_n \to x$ in $d + (d' \circ f)$.

Theorem 3.3. (Wendy Robertson's Open Mapping Theorem)

Let X be a Frechet space, $(Y,\mathcal{J})$ an ultrabarreled linear separated space. Let $f:X \to Y$ be continuous, linear, and onto. Then f is open.

Proof: By example 3.8 it is sufficient to show that f is almost open. Let $\mathcal{V}$ be the topology for Y generated by $\{\overline{f[U]}$: U is a neighborhood of 0 in X$\}$. $\mathcal{V}$ is weaker than $\mathcal{J}$, by definition of ultrabarreled, hence each $\overline{f[U]}$ is a $\mathcal{J}$ neighborhood of 0.

Definition: *Let* $f:X \to Y$. *We call* f *a* KC *map if the image under* f *of every compact set is closed.*

Lemma 3.6. *Let* X *be a locally compact topological space. Let* $f:X \to Y$ *be an almost open* KC *map. Then* f *is open.*

Proof: Let U be a compact neighborhood of $x \in X$. Then $f[U] = \overline{f[U]}$ which is a neighborhood of $f(x)$ in Y.

Corollary: *Let* X *be locally compact,* $f: X \to Y$ *an almost open map with closed graph. Then* f *is open.*

Proof: Closed graph implies KC by [13], §11.1, Fact ii.

Corollary: *Let* X *be locally compact, and let* Y *be a* T_2 *space. If* $f: X \to Y$ *is continuous and almost open,* f *is open.*

Proof: Since f is a continuous map into a T_2 space, f has closed graph.

Theorem 3.4. *Let* $f:X \to Y$ *be one to one. The following are pairwise equivalent.*

i. f *is almost open*

ii. *For every open set* $G \subset Y$, $f^{-1}[\overline{G}] \subset \overline{f^{-1}[G]}$.

iii. *For every closed set* $F \subset Y$, $f^{-1}[F^i] \supset [f^{-1}[F]]^i$.

[*We remark that* ii $\to$ iii $\to$ i *without the assumption that* f *be one to one*].

Proof: i $\to$ ii Let $x \in f^{-1}[\overline{G}]$. Let V be a neighborhood of x. $\overline{f[V]}$ is a neighborhood of $f(x)$, hence it meets G. Thus $f[V]$ meets G. So V meets $f^{-1}[G]$, and $x \in \overline{f^{-1}[G]}$.

ii $\to$ iii. $\tilde{F}$ is open, so $f^{-1}[\overline{\tilde{F}}] \subset \overline{f^{-1}[\tilde{F}]}$. But $f^{-1}[\tilde{\overline{\tilde{F}}}] = \widetilde{f^{-1}[F^i]} = (f^{-1}[F^i])^\sim$, and $\overline{f^{-1}[\tilde{F}]} = (\overline{f^{-1}[F]})^\sim = \widetilde{(f^{-1}[F])^i}$

iii $\rightarrow$ i Let V be a neighborhood of x. Then $x \in V^i \subset (f^{-1}[f[V]])^i \subset (f^{-1}[\overline{f[V]}])^i \subset f^{-1}[[\overline{f[V]}]^i]$. So $f(x) \in (\overline{f[V]})^i$.

Theorem 3.5. _In theorem 3.4 we suppose that f is continuous, the_ i $\Leftrightarrow$ ii$'$ $\Leftrightarrow$ iii$'$ _where_ ii$'$ = ii _with inclusion replaced by equality and similarly for_ iii$'$.

Theorem 3.6. _Let_ $f:X \rightarrow Y$ _be continuous and almost open. If_ X _is of category_ II, _so is_ Y. _If_ X _is first countable and_ Y _is regular,_ Y _is first countable. If_ X _is a paranormed linear topological space and_ Y _is a linear topological space,_ Y _is paranormed._

Proof: If Y is of category I, $Y = \bigcup_{n=1}^{\infty} F_n$ where each F_n is closed and nowhere dense. Then $X = \bigcup_{n=1}^{\infty} f^{-1}[F_n]$, and $(f^{-1}[F_n])^i \subset f^{-1}[F_n^i] = \emptyset$. So X is of category I.

Let $\{U_n\}$ be a basic sequence of neighborhoods of X. $\{\overline{f[U_n]}\}$ is a basic sequence of neighborhoods of f(x).

The last part of the theorem follows from the second part since a linear topological space is first countable if and only if it is paranormed.

We note that even continuous open maps do not necessarily preserve separation. If L is a non-closed linear subspace of a linear topological space X, X/L is not T_2.

§4 Weston's Ordering Relation

Let $\mathcal{J}$ and $\mathcal{J}'$ be topologies for a set X. We say $\mathcal{J} > \mathcal{J}'$ if $i: (X,\mathcal{J}) \to (X,\mathcal{J}')$ is continuous and almost open.

We note that $>$ is transitive, reflexive, and antisymmetric. Also if $\mathcal{J} > \mathcal{J}'$, $\mathcal{J}$ is stronger than $\mathcal{J}'$. However, $>$ is not directed from above or below as we see in the following example.

Example 4.1. Let $\mathcal{J}$ and $\mathcal{J}'$ be T_1 topologies for a set X such that $\mathcal{J}$ has isolated points, and $\mathcal{J}'$ does not. If $\mathcal{J}'' > \mathcal{J}$, then since $\mathcal{J}'' \supset \mathcal{J}$, $\mathcal{J}''$ has isolated points. If $\mathcal{J}'' > \mathcal{J}'$, $\mathcal{J}'$ has isolated points by the remark on page 15.

A similar argument shows that $>$ is not directed from below.

We call $\mathcal{J}$ P W-minimal if for any topology $\mathcal{J}'$ satisfying property P with $\mathcal{J}' < \mathcal{J}$, we have $\mathcal{J} = \mathcal{J}'$.

If $\mathcal{J}$ is P minimal, then $\mathcal{J}$ is P W-minimal.

If $\mathcal{J}$ is P W-minimum, then $\mathcal{J}$ is P minimum.

Remark: A one to one continuous onto almost open map need not preserve disconnection.

Example 4.2. Let X equal the union of the rationals in [0,1] with the irrationals in [1,2]. Let Y = [0,1]. Define $f: X \to Y$ by

$$f(x) = x \quad \text{if } x \text{ is rational}$$
$$= x - 1 \quad \text{if } x \text{ is irrational.}$$

A complete metric space is T_2 W-minimal by theorem 3.1.

A discrete space is W-isolated.

The cofinite topology is T_1 minimum and T_1 W-minimal; and is T_1 W-minimum among topologies having no isolated points.

Theorem 4.1. Let X be a set $(Y, \mathcal{J}_Y)$ a topological space, and $f: X \to Y$ onto. Then X can be given a maximal topology $\mathcal{J}'$ with f almost open. Indeed we can have $\mathcal{J}' \supset f^{-1}[\mathcal{J}_Y]$.

Proof: Let $\mathcal{J} = f^{-1}[\mathcal{J}_Y]$. Let $\mathcal{A} = \{\mathcal{J}_o : f:(X,\mathcal{J}_o) \to Y$ is almost open$\}$. $\mathcal{A} \neq \emptyset$ since $\mathcal{J} \in \mathcal{A}$.

Let $\mathcal{C}$ be a maximal chain in $\mathcal{A}$, and let $\mathcal{J}' = \vee\{\mathcal{J}_o : \mathcal{J}_o \in \mathcal{C}\}$. By lemma 3.5, $f: (X,\mathcal{J}') \to Y$ is almost open.

Corollary: Let $(X,\mathcal{J})$ be a topological space. There exists a maximal topology $\mathcal{J}'$ with $\mathcal{J}' > \mathcal{J}$.

Theorem 4.2. $\mathcal{J}$ is W-maximal for X if and only if every set $S \subset X$ with the property that $S \subset \frac{i}{S}$ is open.

Proof: Necessity - Let S be a subset of X satisfying $S \subset \frac{i}{S}$ with respect to $\mathcal{J}$. Let $\mathcal{J}' = \mathcal{J} \vee \{\emptyset, S, X\}$. I assert that $\mathcal{J}' > \mathcal{J}$. Clearly $\mathcal{J}' \supset \mathcal{J}$. Let G be a $\mathcal{J}'$ neighborhood of x. Then $G \supset V \cap W$ where V is a $\mathcal{J}$ open neighborhood of x and W is either S or X. In $(X,\mathcal{J})$ we have the following possibilities:

Case 1: $W = X$. Then $G = V$, so $\overline{G} \supset V$ and is thus a $\mathcal{J}$ neighborhood of x.

Case 2: $W = S$. Then $\frac{i}{G} \supset \frac{i}{V \cap S} \supset V \cap \frac{i}{S} \supset V \cap S \ni x$.

Thus, $\mathcal{J}' = \mathcal{J}$, and $S \in \mathcal{J}$.

Sufficiency - Let $\mathcal{J}$ have the mentioned property, and let $\mathcal{J}' > \mathcal{J}$. Let $G \in \mathcal{J}'$. Since $i:(x,\mathcal{J}') \to (X,\mathcal{J})$ is almost open, we have in $(X,\mathcal{J})$ that $\overline{G}$ is a neighborhood of each point of G, that is $G \subset \frac{i}{G}$. Therefore G is $\mathcal{J}$ open, and so $\mathcal{J}' = \mathcal{J}$.

Remark: In a W-maximal topology, every dense set is open, hence the space in question is irresolvable (i.e. a set and its complement cannot both be dense). See D. R. Anderson [17].

Lemma 4.1. *Let* $(X,\mathfrak{T})$ *be a Hausdorff space. Let* $x_n \rightarrow x \in X$ *with* $x_n \neq x$ *for all* n. *Let* $F = \{x, x_1, x_2, x_3, \ldots\}$. *Then if* F *has interior,* X *must have an isolated point.*

Proof: Let x_n be interior to F. Let U be a neighborhood of x_n with $x \notin \overline{U}$. Then $U \cap F$ is finite and is a neighborhood of x_n.

Theorem 4.3. *In any* W-*maximal Hausdorff space* $(X,\mathfrak{T})$ *with no isolated points, sequential convergence is trivial.*

Proof: Let $x_n \rightarrow x \in X$ with $x_n \neq x$ for all n. Let $F = \{x, x_1, x_2, \ldots\}$. $\tilde{F} \cup \{x\}$ is not open, hence is not dense. Hence $\tilde{F}$ is not dense, so F has non-empty interior. The result follows from lemma 4.1.

§5 Topological Completeness

Definition: A topological space $(X,\mathfrak{J})$ is called topologically complete if $\mathfrak{J}$ can be given by a complete metric, d.

We note that d may be assumed to be <1.

It is well known that if X is topologically complete, X is an absolute G_δ (i.e. for any metric space Y with $X \subset Y$, X is a G_δ in Y).

Lemma 5.1. ([12], p. 40) A T_2 space cannot be the union of two or more pairwise disjoint, topologically complete, dense subspaces.

Proof: Let $(X,\mathfrak{J})$ be a T_2 space, $\mathfrak{F}$ a family of disjoint topologically complete dense subspaces. For each $S \in \mathfrak{F}$, let d_S be a complete metric for S with $d_S < 1$. We define a metric d on X as follows:

$$d(x,y) = d_S(x,y) \quad \text{if } x, y \in S.$$

$$= 1 \text{ if there is no } S \in \mathfrak{F} \text{ with } x, y \in S.$$

(X,d) is clearly complete since each d_S is a complete metric.

Consider $i: (X,d) \to (X,\mathfrak{J})$. i is continuous, for if $x_n \to x$ in (X,d), $d(x_n,x) < 1$ eventually. Thus, there exists $S \in \mathfrak{F}$ such that x_n, $x \in S$ for sufficiently large n. Then $x_n \to x$ in S, hence in X.

i is almost open, for let U be d neighborhood of $x \in S$. Then U includes a relative S neighborhood of x. S is dense in X, so $\bar{U}$ includes a $\mathfrak{J}$ neighborhood of x.

By theorem 3.1, i is a homeomorphism. This is a contradiction since no S_ϵ is dense in X with respect to d.

Theorem 4.1. (Victor Klee)

A topologically complete topological group is complete.

Proof: Let G be a topologically complete topological group.

Let $\overline{G}$ be the completion of G. G is a subgroup of $\overline{G}$, and is a proper subgroup if G is not complete. In this case $\overline{G}$ is a disjoint union of left cosets of G. This is impossible by lemma 4.1.

Chapter Two
Inductive Limits

§1 Preliminaries

In this chapter we shall make the following blanket assumptions:

i. Φ is a family of linear topological spaces.

ii. E is a fixed linear space.

iii. For each X in Φ there is a linear map u_X: X → E.

iv. $\bigcup\{u_X[X]: X \in \Phi\}$ spans E.

Definition: A vector topology $\mathfrak{J}$ for E will be called a test topology for E if $u_X: X \to (E,\mathfrak{J})$ is continuous for every $X \in \Phi$.

Definition: Let $\mathfrak{J}_u = \vee\{\mathfrak{J} : \mathfrak{J}$ is a test topology for E$\}$.
Let $\mathfrak{J}_i = \vee\{\mathfrak{J} : \mathfrak{J}$ is a locally convex test topology for E$\}$.

We call $\mathfrak{J}_u$ the unrestricted inductive limit for E with respect to Φ.

We call $\mathfrak{J}_i$ the inductive limit for E with respect to Φ .

Lemma 1.1. Each $u_X: X \to (E,\mathfrak{J}_u)$ is continuous; a fortiori each $u_X: X \to (E,\mathfrak{J}_i)$ is continuous.

Proof. Let $x_\delta \to 0 \in X$. $u_X(x_\delta) \to 0$ in E with any test topology on E. Hence, $u_X(x_\delta) \to 0$ in $(E,\mathfrak{J}_u)$ and in $(E,\mathfrak{J}_i)$.

Example 1.1. $\Phi = \{R^n\}$, $E = E^\infty$, $u_{R^n}(z) = (z_1, z_2, \ldots, z_n, 0, 0, 0, \ldots)$

Example 1.2. E = s, Φ is the collection of all FK spaces, each u_X is the inclusion map from X to E.

Example 1.3. $\Phi = \{X\}$, $u: X \to E$ is onto. Then $\mathcal{J}_u$ is the quotient topology for E with respect to u. If X is locally convex, $\mathcal{J}_u = \mathcal{J}_i$.

§2 Inductive Limits of Locally Convex Spaces

Theorem 2.1. <u>Let each member of Φ be locally convex. Let $\mathcal{U} = \{G : G \subset E$, G is convex, balanced, and $u^{-1}[G]$ is a neighborhood of 0 in X for each $X \in \Phi\}$. Then $\mathcal{U}$ is a local base of neighborhoods of 0 for $\mathcal{J}_i$.</u>

Proof: To prove that the members of $\mathcal{U}$ are $\mathcal{J}_i$ neighborhoods of 0, we show that $\mathcal{U}$ generates a test topology [To see that $U \in \mathcal{U}$ is absorbing, let $z \in E$. Then $z = \sum_{i=1}^{n} u_{X_i}(x_i)$. Choose $\epsilon > 0$: $|\alpha| < \epsilon$ implies that $\alpha x_i \in \frac{1}{n} u_{X_i}^{-1}[U]$. Then $\alpha z \in \sum_{i=1}^{n} (\frac{1}{n} U) = U$. The other property of [13], §10.1, Theorem 3 clearly holds; namely for $G \in \mathcal{U}$ there exists $G_1 \in \mathcal{U}$ with $G_1 + G_1 \subset G$, for let $G_1 = \frac{1}{2}G$.

Conversely, any convex, balanced $\mathcal{J}_i$ neighborhood of 0 belongs to $\mathcal{U}$ by Lemma 1.1

Example 2.1. $\Phi = \{R, R\}$, $E = R^2$, $u_1(x) = (x,0)$, $u_2(x) = (0,x)$.

Let $G = \{(0,t) : |t| < 1\} \cup \{(t,0) : |t| < 1\}$. $u_1^{-1}[G]$ and $u_2^{-1}[G]$ are absorbing, but G is not absorbing.

Corollary: <u>An inductive limit of</u> t <u>spaces is a</u> t <u>space</u>.

Proof: Let B be a $\mathcal{J}_i$ barrel in E. Then $u_X^{-1}[B]$ is a barrel in X, hence a neighborhood of 0 in X, for every $X \in \Phi$. By Theorem 2.1, B is a $\mathcal{J}_i$ neighborhood of 0.

Corollary 2: <u>A quotient of a</u> t <u>space is a</u> t <u>space</u>.

§3 Finite Unrestricted Inductive Limits

Theorem 3.1. *Let $\Phi = \{X_1, X_2, \ldots, X_m\}$. Let $\mathcal{U} = \{\sum_{i=1}^{m} u_i[V_i] : V_i$ is a balanced neighborhood of 0 in X_i for each $i.\}$ Then $\mathcal{U}$ is a local base of neighborhoods of 0 for $\mathfrak{J}_u$.*

Proof: Let V be a $\mathfrak{J}_u$ neighborhood of 0. Let W be a balanced $\mathfrak{J}_u$ neighborhood of 0 with $\sum_{i=1}^{m} W \subset V$. Then $V \supset \sum_{i=1}^{m} u_i\{u_i^{-1}[W]\} \in \mathcal{U}$.

The converse follows since $\mathcal{U}$ generates a test topology. [Let $U \in \mathcal{U}$, say $U = \sum_{i=1}^{m} u_i[V_i]$ where each V_i is a balanced neighborhood of 0 in X_i. To see that U is absorbing, let $z \in E$. $z = \sum_{i=1}^{m} u_i(x_i)$. Choose $\epsilon > 0$: $|\alpha| < \epsilon$ implies that $\alpha x_i \in V_i$ for each i. Then for $|\alpha| < \epsilon$, $\alpha z = \sum_{i=1}^{m} u_i(\alpha x_i) \in \sum_{i=1}^{m} u_i[V_i] = U$. So U is absorbing. Next we show that there exists $U_1 \in \mathcal{U} : U_1 + U_1 \subset U$. Choose $W_i \subset X_i$ such that W_i is a balanced neighborhood of 0 in X_i with $W_i + W_i \subset V$. Let $U_1 = \sum_{i=1}^{m} u_i[W_i]$. Then $U_1 + U_1 \subset U$. The other conditions of [13], §10.1, theorem 3 are easily verified].

Corollary: If Φ is a finite collection of locally convex spaces, $\mathfrak{J}_u = \mathfrak{J}_i$.

See [18], pages 131, 133, 146 and [19].

§4 Functions on Inductive Limits: Bornology

Theorem 4.1. (a) *Let X be locally convex for every X in Φ. Let Y be a locally convex linear topological space, and $f: E \to Y$ a linear map.*

<u>The following are pairwise equivalent.</u>

i. f <u>is</u> $\mathfrak{J}_i$ <u>continuous</u>

ii. f <u>is</u> $\mathfrak{J}_u$ <u>continuous</u>

iii. $f \circ u_X$ <u>is continuous for every</u> X <u>in</u> Φ .

(b) <u>Let</u> $\Phi = \{X_1, \ldots, X_m\}$, Y <u>be a linear topological space, and</u> $f: E \to Y$ <u>be a linear map. Then</u> f <u>is</u> $\mathfrak{J}_u$ <u>continuous if and only if</u> $f \circ u_i$ <u>is continuous for</u> $i = 1, 2, \ldots, m$.

Proof: (a) i $\to$ ii $\to$ iii Trivial

iii $\to$ i Let V be a convex, balanced neighborhood of 0 in Y. For every X in Φ , $u_X^{-1}\{f^{-1}[V]\} = (f \circ u_X)^{-1}[V]$ is a neighborhood of 0 in X. Therefore, $f^{-1}[V]$ is a neighborhood of 0 in $(E, \mathfrak{J}_i)$ by Theorem 2.1.

(b) Necessity - Trivial

Sufficiency - Let V be a balanced neighborhood of 0 in Y. Let W be a balanced neighborhood of 0 in Y with $\sum_{i=1}^{m} W \subset V$. Let $U_i = (f \circ u_i)^{-1}[W]$ for $i = 1, \ldots, m$. Then $\sum_{i=1}^{m} u_i[U_i] \subset f^{-1}[V]$. So $f^{-1}[V]$ is a $\mathfrak{J}_u$ neighborhood of 0 in $(E, \mathfrak{J}_u)$. Thus, f is $\mathfrak{J}_u$ continuous.

Corollary 1: If each X in Φ is locally convex, $\mathfrak{J}_i$ and $\mathfrak{J}_u$ have the same dual.

Corollary 2: An inductive limit of bornological spaces is bornological.

Proof: Let E be such an inductive limit, Y a locally convex linear topological space, $f: E \to Y$ linear and bounded.

For each X in $\mathfrak{X}$, $f \circ u_X$ preserves bounded sets, hence is continuous since X is bornological. The result follows from part (a) of the theorem.

Corollary 3: A quotient of a bornological space is bornological.

Theorem 4.2. Any Bornological space $(E,\mathfrak{J})$ is the inductive limit of semi-normed spaces. If E is separated (complete) [separated and complete], the spaces may be taken to be normed (complete) [Banach] spaces.

Proof: For every closed, bounded, balanced, convex set B in E, let E_B be the span of B, and let p_B be the gauge of B.

On E_B, p_B is stronger than $\mathfrak{J}$ since B is bounded. Therefore, if E is separated, p_B is a norm. If E is complete, so is p_B ([5], p. 64, #C).

I assert that $(E,\mathfrak{J})$ is the inductive limit of $\mathfrak{X} = \{E_B\}$, with u_B the inclusion map from E_B into E. [$\mathfrak{J} \subset \mathfrak{J}_i$ since $\mathfrak{J}$ is a test topology for E. Conversely, let V be a $\mathfrak{J}_i$ neighborhood of 0. We shall show that V is a bornivore, hence V is a $\mathfrak{J}$ neighborhood of 0. Let S be a bounded set, and let B equal the convex, balanced, closure of S. Then $u_B^{-1}[V]$ is a neighborhood of 0 in E_B, so it includes $(p_B < \epsilon) = \epsilon B$ for some $\epsilon > 0$. Therefore, $V \supset V \cap E_B = u_B^{-1}[V] \supset \epsilon B \supset \epsilon S$].

Remark: Let $(E,\mathfrak{J})$ be an arbitrary locally convex linear topological space. Let $\mathfrak{J}^b$ be the inductive limit defined in the proof of

Theorem 4.2. Then $\mathfrak{T}^b$ is stronger than $\mathfrak{T}$, and $\mathfrak{T}^b$ is bornological by the second corollary of Theorem 4.1. It is easy to see that $\mathfrak{T}$ and $\mathfrak{T}^b$ have the same bounded sets.

Theorem 4.3. *Associated with each locally convex linear topology $\mathfrak{T}$ on a linear space E is a largest topology $\mathfrak{T}^b$ with the same bounded sets. $\mathfrak{T}^b$ is bornological.*

Remark: It can be proved that $\mathfrak{T}^b = \mathfrak{T}(E,E^b)$ where E^b is the collection of bounded linear functionals on E.

§ 5 Embedded Spaces

In this section we assume that each X in is a linear subspace of a fixed linear space E, and that each u_X is the inclusion map from X to E.

Theorem 5.1 $\mathfrak{J}_u(\mathfrak{J}_i)$ is the largest (locally convex) linear topology for E which induces on each X in Φ a smaller topology than its initial one.

Proof: Each u_X is continuous, and $\mathfrak{J}_u(\mathfrak{J}_i)$ is the supremum of all (locally convex) linear topologies for which this is true.

Example 5.1. Let $E_1 = \bigcap\{X : X \in \Phi\}$ then $\mathfrak{J}_u$ restricted to E_1 is the infimum of $\{\mathfrak{J}_X/E_1 : X \in \Phi\}$.

If Φ is a family of locally convex spaces, these are the same. This generalizes Theorem 2.7 of chapter one.

In particular, if $\mathfrak{J}$ and $\mathfrak{J}'$ are locally convex topologies for a linear space, then $\mathfrak{J} \wedge \mathfrak{J}'$ is an unrestricted inductive limit.

Corollary 5.1. Each X in Φ is continuously embedded in $(E, \mathfrak{J}_u)$ and in $(E, \mathfrak{J}_i)$.

§ 6 Direct Sum

Let $\{X_\alpha : \alpha\epsilon A\}$ be a collection of linear topological spaces. For each $\alpha\epsilon A$, let j^α be the injection of X_α into $\prod_{\alpha\epsilon A} X_\alpha$.

The direct sum of $\{X_\alpha : \alpha\epsilon A\}$, denoted by $\sum_{\alpha\epsilon A} X_\alpha$, is the linear span of $\bigcup_{\alpha\epsilon A}\{j^\alpha X_\alpha\}$.

Example 6.1. Let A be the positive integers with $X_\alpha = R$ for every $\alpha\epsilon A$. Then $\prod_{\alpha\epsilon A} X_\alpha = s$, and $\sum_{\alpha\epsilon A} X_\alpha = R^\infty$.

The (unrestricted) inductive limit topology for $\sum_{\alpha\epsilon A} X_\alpha$ with respect to $\{j^\alpha : \alpha\epsilon A\}$, denoted by $(D_u)D_i$, is called the (unrestricted) direct sum topology.

Remarks: (1) D_u is always larger than the product topology; and if each X_α is locally convex, D_i is larger than the product topology.

(2) If each X_α is separated, D_u is separated. If each X_α is locally convex and separated, D_i is separated.

(3) Each projection $p^\beta : \sum_{\alpha\epsilon A} X_\alpha \to X_\beta$ is D_u continuous, and is D_i continuous if each X_α is locally convex. Also, $p^\alpha \cdot j^\alpha$ is the identity on X_α, and $p^\alpha \cdot j^\beta = 0$ if $\alpha \neq \beta$.

(4) D_u is the largest linear topology which agrees with the initial topology on each X_α; more precisely, D_u is the largest linear topology for $\sum_{\alpha\epsilon A} X_\alpha$ such that each j^α is a homeomorphism. The same is true for D_i if each X_α is locally convex.

__Lemma 6.1. Let__ V __be a product neighborhood of__ 0 __in__ $\sum_{\alpha\in A} X_\alpha$. __Then__ $X^\alpha = p^\alpha[V]$ __for all but finitely many__ $\alpha\in A$.

Proof: $V = W \cap \sum_{\alpha\in A} X_\alpha$ where W is a product neighborhood of 0. $W \supset \bigcap_{i=1}^{n} (p^{\alpha_i})^{-1}[G_i]$, where G_i is a neighborhood of 0 in X_{α_i}. If $\beta \neq \alpha_i$ for $i = 1,\ldots,n$, $W \cap \sum_{\alpha\in A} X_\alpha \supset j^\beta[X^\beta]$. Thus, $p^\beta[V] = X_\beta$.

__Theorem 6.1. Let__ $\{X_\alpha : \alpha\in A\}$ __be an infinite family of non-indiscrete locally convex linear topological spaces. Then__ D_i __is strictly stronger than the product topology.__

Proof: For each $\alpha\in A$ let V_α be a proper, convex, balanced neighborhood of 0 in X_α. Let $V = (\prod_{\alpha\in A} V_\alpha) \cap (\sum_{\alpha\in A} X_\alpha)$. By the preceeding lemma, V is not a product neighborhood of 0 in $\sum_{\alpha\in A} X_\alpha$. V is a D_i neighborhood of 0 by Theorem 2.1 since $(j^\alpha)^{-1} = V_\alpha$.

Robertson and Robertson prove Theorem 6.1, and that for any finite subset $B \subset A$, the product and direct sum topologies agree on $\sum_{\alpha\in B} X_\alpha$.

__Theorem 6.2. Let__ $\{X_\alpha : u_\alpha ; \alpha\in A\}$ __be a collection of locally convex linear topological spaces and linear maps to a linear space__ E. __Let__ $h:(\sum_{\alpha\in A} X_\alpha, D_i) \to E$ __be defined by__ $h(z) = \sum_{\alpha\in A} u_\alpha(p^\alpha(z))$. __Then__ $\mathcal{J}_i$ __is the quotient topology by__ h.

Proof: By property iv of our blanket assumption, h is onto.

Since $h_0 j^\alpha = u_\alpha$ for each α, h is continuous by Theorem 4.1, remembering that $\sum_{\alpha\in A} X_\alpha$ is an inductive limit.

Finally, h is open, for let V be a convex balanced neighborhood of 0 in $\sum_{\alpha\in A} X_\alpha$. For every $\beta\in A$, $u_\beta^{-1}[h[V]] = j_\beta^{-1}[V]$ is a neighborhood of 0 in X_β. Therefore, h[V] is a $\mathfrak{J}_i$ neighborhood of 0 in E by Theorem 2.1.

<u>Corollary:</u> *$\mathfrak{J}_i$ is separated if and only if $h^\perp$ is closed.*

Example 6.2. Let $\mathfrak{J}$ and $\mathfrak{J}'$ be locally convex linear topologies for a linear space X. By the corollary, $\mathfrak{J}_1 \wedge \mathfrak{J}_2$ is separated if and only if $h^\perp$ is closed where $h: (X,\mathfrak{J}_1) \times (X,\mathfrak{J}_2) \to X$ is defined by $h(x,y) = x + y$; that is if and only if $\{(x,y): x = -y$ is closed. So $\mathfrak{J}_1 \wedge \mathfrak{J}_2$ is separated exactly when the identity from $(X,\mathfrak{J}_1)$ to $(X,\mathfrak{J}_2)$ has closed graph.

<u>Theorem 6.3.</u> *Let $X_1, X_2, \ldots, X_n$ be linear topological spaces, and for each $i = 1,\ldots,n$, let u_i be a linear map from X_i to a linear space E. Define $h: \sum_{i=1}^n X_i \to E$ by $h(z) = \sum_{i=1}^n u_i[p^i(z)]$. Then $\mathfrak{J}_u$ is the quotient topology for E by h.*

Proof: As in Theorem 6.2, we see easily that h is onto and continuous. To see that h is open, let $V = \sum_{i=1}^n j^i[V_i]$ be a basic neighborhood of 0 in $\sum_{i=1}^n X_i = \prod_{i=1}^n X_i$, where each V_i is a neighborhood of 0 in X_i. Then $h[V] = \sum_{i=1}^n u_i\, p^i[V] = \sum_{i=1}^n u_i[W_i]$ where each W_i is a neighborhood of 0 in X_i.

Remark: In section 5 of this chapter, we considered the assumption that every space in our collection Φ be embedded in a fixed linear space E. We now answer the obvious question of how much generality is lost by this assumption.

Any collection $\{X_\alpha : \alpha\epsilon A\}$ of linear spaces with the property that wherever two of them overlap, their linear operations agree, may be considered as linear subspaces of a fixed linear space; that is, there exists a linear space E and isomorphisms $U_\alpha : X_\alpha \to E$ such that $U_\alpha(x)$ is independent of α if x is an element of more than one X_α.

We may also assume that $\bigcup_{\alpha\epsilon A} U_\alpha[X_\alpha]$ spans E.

Proof: Let $F = \sum_{\alpha\epsilon A} X_\alpha$. Let $L = \{x\epsilon F : \sum_{\alpha\epsilon A} x_\alpha = 0\}$. Let $E = F/L$.

Define $U_\alpha : X_\alpha \to E$ by $U_\alpha(t) = j^\alpha(t) + L$. We note that $U_\alpha = q \circ j^\alpha$ where q is the quotient map from F to E.

Each U_α is an isomorphism. [U_α is clearly linear. If $U_\alpha(t) = 0$, $j^\alpha(t)\epsilon L$; and since $j^\alpha(t)$ has at most one non-zero coordinate, $j^\alpha(t) = 0$, and thus $t = 0$.]

Finally, if $t\epsilon X_\alpha \cap X_\beta$, $\alpha \neq \beta$, $U_\alpha(t) - U_\beta(t) = q[j^\alpha(t) - j^\beta(t)] = 0$ since $j^\alpha(t) - j^\beta(t)\ \epsilon L$.

Example 6.3. R^∞ is a direct sum of copies of R, so it is bornological with D_i. D_i is larger than the product topology, so R^∞ is a K space (sequence space with continuous coordinates).

(δ^n) is both a Hamel and a Schauder basis, so $(R^\infty)' = s$. Since R^∞ is bornological, it is relatively strong, so $D_i = \tau(R^\infty, s) = \tau(s', s)$.

<u>R^∞ has a closed bornivore without interior</u> (since R^∞ is bornological, every <u>convex</u> bornivore is a neighborhood of 0).

Let $A_n = \left\{ \sum_{i=1}^{n} X_i \delta^i : |X_i| \leq {}^1/n \text{ for } i = 1,2,..n. \right\}$

Let $A = \bigcup_{n=1}^{\infty} An$. A is closed. A is a bornivore, because every bounded set in R^∞ must be finite dimensional. ([9], Proposition 24, page 92). A has empty interior since there exists no absorbing set B with $B+B \subset A$ ([2], chapter I, page 21, # 6).

§ 7 Inductive Limits of Paranormed Spaces

Theorem 7.1. The unrestricted inductive limit of a finite number of paranormed spaces is paranormed.

Proof: Let $X_1, X_2, \ldots, X_n$ be linear topological spaces with paranorms $p_1, p_2, \ldots, p_n$ respectively. For each $i = 1,2,\ldots n$, let $\iota_i : X_i \to E$ be a linear map.

For $z \in E$, we define $q(z) = \inf \left\{ \sum_{i=1}^{n} p_i(x_i) : x_i \in X_i \text{ for } i = 1,2,..n, \text{ and } z = \sum_{i=1}^{n} u_i(x_i) \right\}$.

Before showing that q is a paranorm which gives the unrestricted inductive limit topology for E, we give two examples to show the motivation for the definition of q.

1. Suppose we have just two spaces, (X,p) and (Y,h), which are linear subspaces of E, with u_1 and u_2 inclusion maps. Then $q(z) = \inf \left\{ p(x) + h(y) : x \in X, y \in Y, x+y = z \right\}$. If d, d', d'' are the semi -metrics corresponding to $q, p,$ and h respectively, we have $d(z,w) = q(z-w) = \inf \left\{ d(z,t) + d(t,w) : t \in X \cap Y \right\}$.

2. Let X be a normed space, L a linear subspace, and $u : X \to X/L$ the quotient map. Now $q(z) =$

$$q(x+L) = \inf \left\{ \|t\| : u(t) = x+L \right\}$$
$$= \inf \left\{ \|t\| : t \in x+L \right\}$$

We now show that q is a paranorm.

It is clear that $q(0) = 0$, and that for every $z \in E$, $q(z) \geq 0$ and $q(z) = q(-z)$.

$$q(z+w) = \inf\left\{\sum_{i=1}^{n} p_i(x_i) : \sum_{i=1}^{n} u_i(x_i) = z+w\right\}$$

$$= \inf\left\{\sum_{i=1}^{n} p_i(x_i+y_i) : \sum_{i=1}^{n} u_i(x_i+y_i) = z+w\right\}$$

$$\leq \inf\left\{\sum_{i=1}^{n} p_i(x_i+y_i) : \sum_{i=1}^{n} u_i(x_i) = z, \sum_{i=1}^{n} u_i(y_i) = w\right\}$$

$$\leq \inf\left\{\sum_{i=1}^{n} p_i(x_i) + \sum_{i=1}^{n} p_i(y_i) : \sum_{i=1}^{n} u_i(x_i) = z, \sum_{i=1}^{n} u_i(y_i) = w\right\}$$

$$= \inf\left\{\sum_{i=1}^{n} p_i(x_i) : \sum_{i=1}^{n} u_i(x_i) = z\right\} + \inf\left\{\sum_{i=1}^{n} p_i(y_i) : \sum_{i=1}^{n} u_i(y_i) = w\right\}$$

(Since x_i and y_i can be chosen independently of each other)

$$= q(z) + q(w).$$

We show continuity of multiplication in three steps.

1. Let $q(z^k) \to 0$. Let $t_k \to t$, in K. For each positive integer k, choose $x_i^k \in X_i$ such that $\sum_{i=1}^{n} u_i(x_i^k) = z^k$, and $\sum_{i=1}^{n} p_i x_i^k < q(z^k) + \frac{1}{k}$.

Then for each $i = 1,2,..n$, $x_i^k \to 0$ in X_i. Therefore, $t_k x_i^k \to 0$ in X_i. Thus, $q(t_k x^k) \leq \sum_{i=1}^{n} p_i(t_k x_i^k) \to 0$. So $q(t_k x^k) \to 0$.

2. Let $t_k \to 0$ in K Fix $z \in E$. $z = \sum_{i=1}^{n} u_i(x_i)$ where $x_i \in X_i$ for $i = 1,2,\ldots n$. So $t_k z = \sum_{i=1}^{n} u_i(t_k x_i)$. It follows that $q(t_k z) \leq \sum_{i=1}^{n} p_i(t_k x_i)$ which converges to 0. So $q(t_k z) \to 0$.

3. Finally, let $q(z^k - z) \to 0$, $t_k \to t$ in K. $q(t_k z^k - tz) \leq q[t_k(z_k - z)] + q[(t_k - t)z]$ which converges to 0 by 1 and 2.

We now show that q gives the $\mathfrak{J}_u$ topology for E. Since for $1 \leq i \leq n$, $u_i(x_i) = u_1(0) + \ldots + u_i(x_i) + \ldots + u_n(0)$, we see that $q[u_i(x_i)] \leq p_i(x_i)$. Thus, $u_i : X_i \to (E,q)$ is continuous, and so u is stronger than q.

Conversely, it suffices to consider sequences, since (E,q) is first countable. Let $q(z^k) \to 0$. For each integer k, and $i = 1,2,\ldots n$, choose x_i^k with $\sum_{i=1}^{n} u_i(x_i^k) = z^k$ and $\sum_{i=1}^{n} p_i(x_i^k) < q(z^k) + \frac{1}{k}$. Then $x_i^k \to 0$ in X_i for $i = 1,2,\ldots n$. So $z^k = \sum_{i=1}^{n} u_i(x_i^k) \to 0$ in $\mathfrak{J}_u$ since each u_i is continuous from X_i to $(E, \mathfrak{J}_u)$. Thus, q is stronger than $\mathfrak{J}_u$.

Theorem 7.2. <u>If in theorem 7.1, each X_i is locally convex, so is $\mathfrak{J}u$. If each X_i is semi-normed, so is $\mathfrak{J}u$.</u>

Proof: The first part of the theorem follows from theorem 3.1.

Suppose each p_i is a semi-norm. Let t be a non-zero scalar. $q(tz) = \inf \left\{ \sum_{i=1}^{n} p_i(x_i) : x_i \epsilon X_i \text{ for } i = 1,2,\ldots n, \; tz = \sum_{i=1}^{n} u_i(x_i) \right\} = |t| \inf \left\{ \sum_{i=1}^{n} p_i(\frac{xi}{t}) : \sum_{i=1}^{n} u_i(\frac{xi}{t}) = z \right\}$

$$= |t| \inf \left\{ \sum_{i=1}^{n} p_i(y_i) : \sum_{i=1}^{n} u_i(y_i) = z \right\}$$

$$= |t| \; q\,(z)$$

Theorem 7.3. <u>If in theorem 7.1 each X_i is complete, u is complete.</u>

Proof: Let $\{z^k\}$ be a Cauchy sequence in (E,q). We may assume that $\sum_{k=1}^{\infty} q(z^k - z^{k-1}) < \infty$. Let $z^o = 0$. For each positive integer k, and $i = 1,2,\ldots n$, choose $x_i{}^k \epsilon X_i$ with

$$\sum_{i=1}^{n} u_i(x_i{}^k) = z^k - z^{k-1}, \quad \sum_{i=1}^{n} p_i(x_i{}^k) < q(z^k - z^{k-1}) + \tfrac{1}{k}2.$$

Let $a_i = \sum_{k=1}^{\infty} x_i^{\,k}$ (note that this series is absolutely convergent).

Then $z^k \to \sum_{i=1}^{n} u_i(a_i)$ since $z^k = \sum_{\Omega=1}^{k} (z^{\Omega} - z^{\Omega-1})$

$$= \sum_{\Omega=1}^{k} \sum_{i=1}^{n} u_i(x_i^{\,\Omega}) = \sum_{i=1}^{n} \sum_{\Omega=1}^{k} u_i(x_i^{\,\Omega}) \to \sum_{i=1}^{n} u_i(a_i).$$

Remark: In general the inductive limit of complete spaces is not complete, even if $n = 1$. Namely, there exists a locally convex linear separated space x, and a closed linear subspace L of x such that x/L is not complete (see [20], [5] page 195, #D).

However, a strict inductive limit of complete spaces is complete ([9], page 128, proposition 3).

Example 7.1. Let $(B, \| \,\|)$ be a Banach space, X and Y disjoint closed linear subspaces. Let $E = X + Y$, and let $u : x \to E$, $v : y \to E$ be inclusion maps. The unrestricted inductive limit topology for E is given by the semi-norm $q(z) = \{\|x\| + \|y\| : x+y = z\}$ (We do not need to take the infimum since we have unique representation in this case). q is clearly a norm, and is complete by Theorem 7.3.

$\|x+y\| \le \|x\| + \|y\| = q(x+y)$, so q and $\|\cdot\|$ are equivalent if and only if E is $\|\cdot\|$- closed in X.

There are examples in which E is not $\| \ \|$ closed in X, so in such cases we have two different norms for a linear space E which agree on two complementary closed linear subspaces.

§ 8 Separation of the Inductive Limit

Theorem 8.1 Let X and Y be linear topological spaces, E a linear space, and $u : X \to E$ and $v : Y \to E$ linear maps. If E has a Hausdorff topology $\mathcal{J}$ making u and v continuous, $\mathcal{J}_u$ is separated.

Proof: It is sufficient to show that $h^{\perp}$ is closed in $X \times Y$, where $h(x,y) = u(x) + v(y)$. Let (x_δ, y_δ) be a net of points in $h^{\perp}$ converging to (x,y) in $X \times Y$. Since $-y_\delta \to -y$ in Y and v is linear, we have $-v(y_\delta) = v(-y_\delta) \to v(-y) = -v(y)$ in $\mathcal{J}$. Since $u(x_\delta) = -v(y_\delta)$ for each δ, $u(x_\delta)$ converges to both $u(x)$ and $-v(y)$ in the Hausdorff topology $\mathcal{J}$. Thus, $u(x) + v(y) = 0$ and $(x,y) \in h^{\perp}$.

The following example shows that it is impossible to extend the above theorem to an arbitrary number of spaces.

Example 8.1 Let L be a linear space, and let $\mathcal{J}$ and $\mathcal{J}'$ be separated linear topologies for L such that $\mathcal{J} \cap \mathcal{J}'$ is not Hausdorff (see for example 2.4 of chapter one). We define $X = (L, \mathcal{J})$, $Y = (L, \mathcal{J})$, $Z = (L, \mathcal{J}')$,

$E = L \times L$, $u(x) = (x,0)$, $v(x) = (0,x)$, and $w(x) = (x,x)$.

The topology which agrees with $\mathcal{J}$ on $u[L]$ and $v[L]$, agrees with $\mathcal{J}'$ on $w[L]$, and is discrete elsewhere is a Hausdorff topology making u,v, and w continuous.

Next, $w : (L, \mathcal{J}) \to (E, \mathcal{J}_u)$ is continuous, for let U be a neighborhood of 0 in $(E, \mathcal{J}_u)$. Let V be a neighborhood of 0 in $(E, \mathcal{J}_u)$ with $V + V \subset U$. $N = (u^{-1}[V]) \cap (v^{-1}[V])$ is a $\mathcal{J}$ neighborhood of 0 in L, and $w[N] \subset U$. [Let $x \in N$. $w(x) = (x,x) = (x,0) + (0,x) \in V + V \subset U$].

If $\mathcal{J}_u$ were separated, $w^{-1}(\mathcal{J}_u)$ would be Hausdorff since w is one-to-one. Since $w^{-1}(\mathcal{J}_u)$ is weaker than $\mathcal{J}$ and $\mathcal{J}'$, this is impossible.

<u>Theorem 8.2</u> Let X,Y, and Z be linear topological spaces, E a linear space, and $u : X \to E$, $v : Y \to E$, and $w : Z \to E$ linear maps. If E has a topology $\mathcal{J}$ with closed graph addition making u,v, and w continuous, $\mathcal{J}_u$ is separated.

Proof: Let $(x_\delta, y_\delta, z_\delta)$ be a net in $h^\perp$ converging to (x,y,z) in $(X \times Y \times Z)$. Then $u(x_\delta) + v(y_\delta) = w(-z_\delta) \to w(-z)$. Since $u(x_\delta) \to u(x)$, and $v(y_\delta) \to v(y)$, $u(x) + v(y) = w(-z)$. Thus, $h^\perp$ is closed.

<u>Theorem 8.3</u> Let X, Y, and Z be linear topological spaces, E a linear space, and $u : X \to E$, $v : Y \to E$, and $w : Z \to E$ linear maps. If E has a topology $\mathcal{T}$ with continuous addition making u, v, and w continuous, $\mathcal{T}_u$ is stronger than $\mathcal{T}$.

<u>Proof</u>: $h(x,y,z) = u(x) + v(y) + w(z)$, so h is continuous.

<u>Corollary</u> If in theorem 8.3 $\mathcal{T}$ is Hausdorff, then $\mathcal{T}_u$ is separated.

Chapter Three

Some Properties of Linear Topological Spaces

§ 1 Barreled Ideas in Non-Locally Convex Spaces

Definition. A linear topological space is called a barreled space (t space) if every closed, convex, absorbing set has interior.

Definition. A linear topological space is called W-barreled if every closed absorbing set has interior.

Definition. A linear topology $\mathcal{J}$ is called ultra-barreled if every linear topology with a local base of $\mathcal{J}$ closed neighborhoods of 0 is smaller than $\mathcal{J}$.

Proposition. Let $(X,\mathcal{J})$ be a W-barreled linear topological space. Then $\mathcal{J}$ is ultra-barreled.

Proof: Let $\mathcal{J}'$ be a linear topology with a local base of $\mathcal{J}$ closed neighborhoods of 0. Each $\mathcal{J}$ closed $\mathcal{J}'$ neighborhood of 0 has interior with respect to $\mathcal{J}$, so $\mathcal{J}$ is stronger than $\mathcal{J}'$.

For locally convex spaces, category II implies W-barreled which implies ultra-barreled which implies barreled. Barreled does not imply ultra-barreled, and ultra-barreled does not imply W-barreled.

Example 1.1. A locally convex space which is ultra-barreled but not W-barreled.

Let $X = R^\infty$ with the direct sum topology, which is the largest linear topology making the injection maps continuous. This is also the largest linear topology for R^∞, so X is clearly ultra-barreled.

X is locally convex ([5], page 53, # I).

X is category I because it has countable Hamel dimension.

By example 6.3 of chapter two, X has a closed bornivore without interior, hence is not W-barreled.

Example 1.2. A linear topological space which is barreled but not ultra-barreled. (See [11]).

Let $(X,P) = \ell^{1/2}$ with the paranorm

$$!x! = \sum |x_n|^{1/2}.$$ (X,P) is a Frechet space.

$(X,N) = \ell^{1/2}$ with the norm

$$\|x\| = \sum |x_n|.$$

Clearly P is strictly stronger than N, since P is not locally convex. (X,N) is not ultra-barreled, for if it were, the identity map from (X,N) to (X,P) would be open by Theorem 3.3 of chapter one.

(X,N) is barreled, for let B be a barrel in (X,N). Then B is a P neighborhood of 0 since B is P closed and P is category II, hence W-barreled. The result follows when we show that every convex P neighborhood of 0 is an N neighborhood of 0.

Let T be the linear topology generated by the convex P neighborhoods of 0 (the associated locally convex topology of P). We show that N is equal to T.

N is weaker than T.

To prove N is stronger than T we observe that N is locally convex and metric, hence relatively strong ([13], § 13.5, # 6; § 12.3, corollary 1; § 10.5, example 4).

Thus, we need only show that

$$(\ell^{1/2},N)' = (\ell^{1/2},T)', \quad \text{that is}$$

$$(\ell^{1/2},N)' = (\ell^{1/2},P)'.$$

This follows since $\{\delta^n\}$ is a Schauder basis for $\ell^{1/2}$ with both N and P. So every linear functional f on $\ell^{1/2}$ has the representation

$$f(x) = \sum_{k=1}^{\infty} t_k x_k \quad \text{for every } x \text{ in } \ell^{1/2}.$$

f is continuous if and only if $t \in m$ for either P or N.

<u>Theorem</u> 1.1. If an ultra barreled space $(X, \mathcal{J})$ is not locally convex, there is no locally convex topology larger than $\mathcal{J}$.

Proof. Let $\mathcal{J}$ be ultra-barreled and $\mathcal{J}'$ a locally convex topology for X which is stronger than $\mathcal{J}$. Let $\mathcal{J}''$ be the linear topology generated by $\mathcal{U}$, the set of convex $\mathcal{J}'$ neighborhoods of 0 which are $\mathcal{J}$ closed. $\mathcal{J}$ is clearly larger than $\mathcal{J}''$. Also, $\mathcal{J}$ is weaker than $\mathcal{J}''$. [Let U be a $\mathcal{J}$ closed $\mathcal{J}$ neighborhood of 0. Then U is a $\mathcal{J}'$ neighborhood of 0. Therefore, U includes a convex $\mathcal{J}'$ neighborhood V of 0. Thus, U includes $cl_{\mathcal{J}} V$, a $\mathcal{J}''$ neighborhood of 0]. Thus, $\mathcal{J}$ is locally convex.

Example 1.3. A non locally convex topology which allows a larger locally convex topology.

Let $X = R^{\infty}$ with the topology induced by the paranorm of $\ell^{1/2}$.

X is not locally convex, for let G be a convex neighborhood of 0. Then $G \supset (\| \ \| \leq \epsilon)$ for some $\epsilon > 0$. Thus, $\epsilon\, \delta^n$ belongs to G, so

$$\frac{1}{n} \sum_{k=1}^{n} \epsilon\, \delta^k \text{ belongs to } G.$$

But $$\left\| \frac{1}{n} \sum_{k=1}^{n} \epsilon\, \delta^k \right\| > 1, \quad \text{for sufficiently large } n.$$

The direct sum topology is a locally convex topology for X larger than the one given by the paranorm of $\ell^{1/2}$.

<u>Corollary</u>. Let $(X, \mathcal{J})$ be a separated ultra-barreled space with $(X, \mathcal{J})' = \{0\}$. Then no locally convex topology is comparable with $\mathcal{J}$.

<u>Theorem</u> 1.2. A pointwise bounded family Φ of continuous linear maps from an ultra-barreled space X to any linear topological space Y is equicontinuous.

[Compare with [13], § 12.3, Th. 4].

<u>Proof</u>. Let $\mathcal{U}$ be the set of closed neighborhoods of 0 in Y. Then $\Phi^{-1}[U] = \bigcap_{f \in \Phi} f^{-1}[U]$ is closed for each $U \in \mathcal{U}$, and the set of all such sets generates a linear topology. This topology is weaker than the original topology of X, so each $\Phi^{-1}[U]$ is a neighborhood of 0.

<u>Corollary</u>. The Banach-Steinhaus Closure Theorem ([13], § 12.3, Th. 5).

Let $\{f_n\}$ be a pointwise convergent sequence of continuous linear functionals from an ultra-barreled space X to a linear topological space Y. Let $f(x) = \lim_{n \to \infty} f_n(x)$. Then f is continuous.

<u>Example</u> 1.4. Let $\mathcal{J}$ and $\mathcal{J}'$ be linear topologies for X such that $\mathcal{J}$ is ultra-barreled and $\mathcal{J}'$ has a Schauder basis $\{b^n\}$ with biorthogonal functions g_n in $(X, \mathcal{J})'$. Then $\mathcal{J}$ is stronger than $\mathcal{J}'$.

<u>Proof</u>. Let $f_n : (X, \mathcal{J}) \to (X, \mathcal{J}')$ be defined by

$$f_n(x) = \sum_{k=1}^{n} g_k(x)b^k.$$

Each f_n is continuous, and f_n converges pointwise to the identity map. By the Banach-Steinhaus closure theorem, the identity is continuous.

<u>Theorem</u> 1.3. A linear map with closed graph from an ultra-barreled space X to a complete paranormed space Y is continuous.

<u>Proof</u>. $\{\overline{f^{-1}[U]} : U \text{ is a neighborhood of } 0 \text{ in } Y\}$ is a base for a vector topology. Therefore, each $\overline{f^{-1}[U]}$ is a neighborhood of 0 in X. The conclusion follows from [5], Theorem 11.1.

<u>Remark</u>. In the definition of ultra-barreled, it is sufficient to assume $\mathcal{J}$ maximal.

<u>Proof</u>. Let $\mathcal{J}'$ be a vector topology with a base of $\mathcal{J}$ closed sets. $\mathcal{J} \vee \mathcal{J}'$ is stronger than $\mathcal{J}$, and has a local base of $\mathcal{J}$ closed sets. Thus $\mathcal{J} \vee \mathcal{J}' = \mathcal{J}$, and $\mathcal{J}$ is stronger than $\mathcal{J}'$.

§ 2 Sequential Properties, Bornology

The interested reader is referred to articles by R. M. Dudley, Transactions of the American Mathematical Society, 1964; S. P. Franklin, Duke Mathematical Journal, Volume 25, 1958; and Seth Warner, Duke Mathematical Journal, Volume 25, 1958.

We begin this section with several basic definitions.

A topology is called sequential if every sequentially closed set is closed.

A set G in a topological space is called sequentially open if for every sequence $\{x_n\}$ converging to a point x in G, x_n belongs to G eventually.

A set G in a linear topological space is called a sequential neighborhood of 0 if every sequence converging to 0 belongs to G eventually.

A linear topology is called C-sequential if every convex sequential neighborhood of 0 is a neighborhood of 0.

A linear topology is called N-sequential if every sequential neighborhood of 0 is a neighborhood of 0.

A linear topological space is called semi-bornological if every bounded linear functional is continuous.

A linear topological space X is called bornological if every bounded linear function from X to any locally convex space is continuous.

A linear topological space is called a Mazur space if every sequentially continuous linear functional is continuous.

The following lemma is an easy consequence of the definitions.

Lemma 2.1 A set F in a topological space is sequentially closed if and only if $\tilde{F}$ is sequentially open.

Theorem 2.1 A topological space is sequential if and only if every sequentially open set is open.

Theorem 2.2 A linear topological space is N-sequential if and only if for any subset S, *the sequential closure of* S *is equal to the closure of* S.

Proof: Sufficiency - Suppose U is not a neighborhood of 0. Then $0 \in \overline{\tilde{U}}$. Therefore, there exists a sequence $\{x_n\}$ in $\tilde{U}$ with x_n converging to 0. Hence, U is not a sequential neighborhood of 0.

Necessity - Let $x \in \bar{S} - S$. Then $\tilde{S}$ is not a neighborhood of x, hence is not a sequential neighborhood of x. So there exists a sequence $\{x_n\}$ converging to x which is frequently in S. Then some subsequence of $\{x_n\}$ is in S and converges to x.

Theorem 2.3 For locally convex spaces we have the implications represented in the following diagram:

Paranormed $\Longrightarrow$ N-sequential $\Longrightarrow$ Sequential

$\Downarrow$

Bornological $\Longrightarrow$ C-sequential $\Longrightarrow$ Semi-bornological $\Longrightarrow$ Mazur

Lemma 2.2 A sequential neighborhood of 0 is a bornivore.

Proof: Let U be a sequential neighborhood of 0, and let B be a set not absorbed by U. Then for every integer n, $B \not\subset n\,U$. Choose $x_n \in B - n\,U$. $\frac{x_n}{n} \notin U$. Thus, $\frac{x_n}{n}$ does not converge to 0. Thus, B cannot be bounded.

Corollary Every bornological space is C-sequential.

Lemma 2.3 A sequentially continuous linear functional is bounded.

Proof: Let f be a sequentially continuous linear functional, and let B be a bounded set. $f^{-1}[N(0,1)]$ is a sequential neighborhood of 0, hence a bornivore. Thus $f^{-1}[N(0,1)]$ includes ϵB for some $\epsilon > 0$, that is $|f(x)| < \frac{1}{\epsilon}$ for all $x \in B$.

Corollary Every semi-bornological space is a Mazur space.

Example 2.1 A space which is semi-bornological but not C-sequential.

Let $X = \ell$ with the weak topology.

Let f be a bounded linear functional on X. Then f is a bounded linear functional on ℓ with the normed topology, hence is norm continuous. So f is weakly continuous.

$N(0,1)$ is not a weak neighborhood of 0. But if x_n converges weakly to 0, x_n converges strongly to 0; and therefore x_n belongs to $N(0,1)$ eventually.

<u>Example</u> 2.2 A space which is bornological but not N-sequential.

Let $X = R^{\infty}$ with the strongest linear topology. By example 6.3 of chapter two, X is bornological and has a bornivore, A, with empty interior.

A is a sequential neighborhood of 0, for let $x_n \to 0$. Then $\{x_n\}$ is finite dimensional (If not, there exists $f \in X^{\#}$ with $f(x_n) = 1$ for infinitely many n. But this is impossible since $E^{\#} = E'$).

In $\text{span}\{x_n\} = S$, $S \cap A$ is a bornivore, hence a neighborhood of 0. So x_n belongs to $S \cap A$ eventually. Thus, x_n belongs to A eventually, and A is a sequential neighborhood of 0.

An example of a linear topological space which is not a Mazur space may be found in the Mathematical Reviews, October 1963, # 4160.

$\ell^{1/2}$ with the topology given by $\|x\| = \sum |x_n|$ is an example of a normed barreled space of category I. Another example may be found in [2], chapter V, page 158, # 10e.

Lemma 2.4 Let $(E, \mathfrak{J})$ be a linear topological space. If E has a compatible bornological topology, $\mathfrak{J}_b$, E is semi-bornological.

Proof: $(E, \mathfrak{J})' = (E, \mathfrak{J}_b)' = \{\mathfrak{J}_b$ bounded linear functionals$\} = \{\mathfrak{J}$ bounded linear functionals$\}$.

Example 2.3 The dual of a non-reflexive Banach space with the weak* topology is not semi-bornological.

Proof: Let $f \in X^{**} - \hat{X}$. Then f is not weak* continuous. However, f is weak* bounded, for if $S \subset X^*$ is weak* bounded, S is norm bounded by the Uniform Boundedness Principle, and f is bounded on S.

Remark: A linear topological space is bornological if and only if it is semi-bornological and relatively strong.

Proof: Necessity is well known.

Conversely we have $X' = (X^b)'$, and hence $X = X^b$.

Theorem 2.4 *Every N-sequential linear topological space* X *is sequential. The converse holds if for every sequential neighborhood* V *of* 0, *there exists a sequential neighborhood* U *of* 0 *with* $U + U \subset V$.

Proof: The first statement is clear. To prove the second statement, let U be a sequential neighborhood of 0. Let $V = \{x : U$ is a sequential neighborhood of $x\}$. Then $0 \in V$. Moreover, V is sequentially open, for let $x \in V$. Let W be a sequential neighborhood of 0 with $W + W \subset U - x$. Let $\{x_n\}$ be a sequence converging to x. Then $x_n - x \to 0$, hence $x_n - x$ belongs to W eventually, that is x_n belongs to $x + W$ eventually. The result follows when we show that $x + W \subset V$. Let $w \in W$. We wish to show that U is a sequential neighborhood of $x + w$. Let $y_n \to x + w$. Then $y_n - x - w \to 0$, hence $y_n - x - w$ belongs to W eventually. So y_n belongs to $x + w + W \subset x + W + W \subset U$ eventually.

Since X is sequential, V is open and thus U is a neighborhood of 0.

Theorem 2.5 _Let_ X _be a pseudonorm space. The following are pairwise equivalent._

i. $N(0,1)$ is a sequential neighborhood of 0 in the weak topology.

ii. $C(0,1)$ is weakly sequentially closed.

iii. Weak and norm convergence of sequences coincide.

iv. 0 is not a weak sequential limit point of $C(0,1)$.

Proof: i implies ii - Suppose $C(0,1)$ is not weakly sequentially closed, say x_n converges weakly to x and that $\|x_n\| = 1$, $\|x\| \neq 1$. Then $\frac{x_n - x}{\|x_n - x\|} \to 0$ weakly. But $\frac{x_n - x}{\|x_n - x\|} \notin N(0,1)$. Therefore, $N(0,1)$ is not a weak sequential neighborhood of 0.

ii implies iii.- Suppose iii is false, say $x_n \to x$ weakly but x_n does not converge to x in the pseudonorm topology. Then $\frac{x_n - x}{\|x_n - x\|}$, or a suitable subsequence converges weakly to 0.

iii implies iv.- Trivial

iv implies i.- Suppose $N(0,1)$ is not a weak sequential neighborhood of 0, say x_n converges weakly to 0 with $\|x_n\| \geq 1$. Then $\frac{x_n}{\|x_n\|}$ converges weakly to 0 and is an element of $C(0,1)$.

We next present a new proof of a well known result.

Remark: A locally convex separated space X with a totally bounded neighborhood of 0 must be finite dimensional.

Proof: It is well known that such a space X must be a normed space. If C, the unit circumference of X, is totally bounded, it may be covered by a finite number of translates of the closed disc about 0 of radius $1/2$, say $C \subset \bigcup_{i=1}^{n} S_i$. Each S_i is convex and norm closed, hence weakly closed. Since $0 \notin \bigcup_{i=1}^{n} S_i$, 0 is not a member of the weak closure of C. Hence, X must be finite dimensional.

<u>Definition</u>: A linear topological space X is called braked if for every sequence $\{x_n\}$ in X converging to 0, there exists a sequence $\{k_n\}$ of real numbers diverging monotonically to ∞ with $k_n x_n \to 0$.

<u>Example</u> 2.4 Every paranormed space is braked.

<u>Example</u> 2.5 The dual of a Banach space with the weak* topology is not braked.

<u>Proof</u>: Choose $f_n \to 0$ in the weak* topology with $\|f_n\| = 1$ for each n. Let k_n be a sequence of scalars diverging monotonically to infinity. Then $\|k_n f_n\| \to \infty$, so $k_n f_n$ cannot converge to 0 in the weak* topology by the Uniform Boundedness Principle.

<u>Lemma</u> 2.5 A bounded linear functional on a braked space is sequentially continuous.

<u>Proof</u>: Let f be a linear functional which is not sequentially continuous. Choose a sequence $\{x_n\}$ with $x_n \to 0$, $f(x_n) \not\to 0$. Let $\{k_n\}$ be a sequence of scalars with $k_n \uparrow \infty$. $k_n x_n$ cannot converge to 0 since $f(k_n x_n)$ is unbounded.

<u>Corollary</u> A braked Mazur space is semi-bornological.

<u>Theorem</u> 2.6 <u>Let</u> X <u>be an infinite dimensional linear space with the topology</u> $\sigma(X,X^{\#})$. <u>Then every norm</u> p <u>on</u> X <u>is sequentially continuous, and discontinuous</u>.

Proof: Let $x_n \to 0$. $\{x_n\}$ is finite dimensional, so $\text{span}\{x_n\}$ is a finite dimensional separated space. Then p is continuous on $\text{span}\{x_n\}$, and so $p(x_n) \to 0$.

Corollary 1. $\sigma(X,X^{\#})$ and $\mathcal{T}(X,X^{\#})$ have the same convergent sequences.

Corollary 2. $\sigma(X,X^{\#})$ is sequentially maximum among locally convex topologies.

Corollary 3. The identity map from $(X, \sigma(X,X^{\#}))$ to $(X, \mathcal{T}(X,X^{\#}))$ is sequentially continuous and not continuous.

Theorem 2.7 *Let A be a convex body (convex set with interior) in a linear topological space. Then the sequential closure of A is equal to the closure of A.*

Proof: We may assume that $0 \in A^i$. Let p be the gauge of A. Then the sequential closure of $A \subset \bar{A} = (p \leq 1)$. But $(p \leq 1) \subset$ sequential closure of A, for if $p(x) \leq 1$, $p(\frac{n}{n+1} x) < 1$, and so $\frac{n}{n+1} x \in A$.

Example 2.6 Let X be a Banach space. Then X' is a sequentially closed subset of $X^{\#}$ with the weak* topology by the Banach-Steinhaus Closure Theorem, but is dense in $X^{\#}$([5], Theorem 16.5).

<u>Definition</u>: An mk space is a topological space with the property that a subset S is closed if $S \cap K$ is closed for every compact metrizable subset K.

<u>Lemma</u> 2.6 Let S be a subset of a Hausdorff space X. Then S is sequentially closed if and only if $S \cap K$ is closed for any compact metrizable subset K.

<u>Proof</u>: Necessity - Let K be a compact metrizable subset of X. Let $x \in \overline{S \cap K}$. Then $x \in \bar{K} = K$, so x belongs to the closure in K of $S \cap K$. Hence, there exists a sequence $\{s_n\}$ of points of $S \cap K$ converging to x. In particular, each s_n belongs to S; so since S is sequentially closed, $x \in S$. Thus, $x \in S \cap K$.

Sufficiency - Let $\{s_n\}$ be a sequence of points in S converging to a point $x \in X$. Let $K = \{x, s_1, s_2, s_3, \cdots\}$. K is homeomorphic to the one-point compactification of the integers, hence is compact and metrizable. By hypothesis, $S \cap K$ is closed. But $x \in \overline{S \cap K}$, so $x \in S \cap K$. In particular, $x \in S$.

<u>Corollary</u> A Hausdorff space is sequential if and only if it is an mk space.

<u>Example</u> 2.7 Let B be a separable Banach space. Let X be the dual of B with the weak* topology. Then X is "almost" an mk space in the sense that a <u>convex</u> set S is closed if and only if $S \cap K$ is closed for every

compact metrizable subset K. In fact, for such a set S, the following three conditions are pairwise equivalent:

1. S is sequentially closed.

2. $S \cap K$ is closed for every compact metrizable subset K.

3. $S \cap D_n$ is closed for every positive integer n (where D_n is the closed disc of radius n).

Proof: 1 implies 2 is clear.

2 implies 3 since D_n is compact and metrizable.

3 implies 1 - Let $\{s_n\}$ be a sequence of points of S converging to a point x in the weak* topology. Then $\{x, s_1, s_2, s_3, \cdots\}$ is weak* compact, hence weak* bounded, hence norm bounded, and is thus a subset of some D_n. Since D_n with the weak* topology is compact and metrizable, $x \in S$.

Remark: In the above example, for any set S which is either convex or bounded; if S is sequentially closed, S is closed.

Proof: If S is convex and sequentially closed, $S \cap D_1$ is closed. By the Krein-Smulian Theorem S is closed.

If S is bounded and sequentially closed, $S \subset D_n$ for some n, and D_n is metrizable. So S is closed.

Example 2.8 In Example 2.7, X need not be N-sequential, that is the assumption that S be convex may not be ommitted.

Let S be a non-norming total subspace of X ([13], § 7.2, # 24). There exists $x \in \bar{S}$ which is not the limit of any norm bounded net in S (see for example the thesis of D. R. Kerr), a fortiori not the limit of any sequence in S.

Chapter 4

FH and BH Spaces

§ 1 The Lattice of FH Spaces

Definition. A topological space X is said to be continuously embedded in a topological space Y if $X \subset Y$ and the inclusion map is continuous.

Let H be a fixed linear space with a Hausdorff topology in which addition is continuous.

Definition. An FH space is a Frechet space which is continuously embedded in H.

As an easy consequence of the Closed Graph Theorem, we have the fact that if $X \subset Y \subset H$ and X and Y are FH spaces, then X is an FY space. In particular, a linear subspace of H has at most one FH topology. (See [13], § 11.3).

Let $\mathfrak{F}_H$ be the collection of all FH spaces. We have

Theorem 1.1. *$\mathfrak{F}_H$ is a lattice. $X \vee Y$ is equal to $X + Y$ with the unrestricted inductive limit topology with respect to the inclusion maps. $X \wedge Y$ is equal to $X \cap Y$ with the supremum of the topologies of X and Y.*

<u>Proof</u>: Theorems 7.1 and 7.3 of chapter two. [13], § 11.3, Theorem 3.

<u>Remark</u>: $X \wedge Y$ exists without the assumption of continuous addition on H. $X \vee Y$ need not exist, however, in the sense that it may not be continuously embedded in H. As an example let X be the x-axis, Y the y-axis, and let H be the plane with the ordinary topology on the axes and the discrete topology elsewhere. $X + Y$ is the ordinary topology for the plane which is not stronger than that of H.

<u>Theorem</u> 1.2. <u>Let $\{X_n\}$ be a sequence of FH spaces. Then $\{X_n\}$ has an infimum which is an FH space.</u>

<u>Proof</u>: Let X be $\bigcap_{n=1}^{\infty} X_n$ with the topology $\bigvee_{n=1}^{\infty} \mathcal{J}_{X_n}$.

That X has the required properties is proved in [13], § 11.3, Theorem 3.

An important special case is that in which $H = s$, the space of all sequences of complex numbers. In this case, an FH space is called an FK space.

In the following examples we concern ourselves with the existence of a supremum of the sequence of FK spaces ℓ^n.

<u>Example</u> 1.1. Let $X = \bigcup_{n=1}^{\infty} \ell^n$. By [13], § 11.3, Corollary 6, X has no FK topology.

Next, we see that c_o is not the sup of $\{\ell^n\}$, for there exists an FK space S such that for every integer n, $\ell^n \subset S \subset c_o$ and $S \neq c_o$.

Let $u \in \ell^{1+\frac{1}{n}} - \ell$ for every n (for example $u = \{\frac{1}{k}\}$). Let $u^\beta = \{y : \sum_{i=1}^{\infty} y_i u_i$ is convergent$\}$. For every integer n, $\ell^n \subset u^\beta$ [Let $y \in \ell^n$ with $n > 1$. $\sum_{i=1}^{\infty} |y_i u_i| \leq \sum_{i=1}^{\infty} |y_i|^n \sum_{i=1}^{\infty} |u_i|^{1 + \frac{1}{n-1}}$ by Holder's Inequality. For $n = 1$, we have $\ell^1 \subset \ell^2$].

Let $S = u^\beta \cap c_o$. $S \neq c_o$, for if $\sum_{i=1}^{\infty} u_i \epsilon_i$ is convergent for every ϵ in c_o, u would be in ℓ by the Uniform Boundedness Principle.

u^β is an FK space by [13], Page 227, Lemma 1, hence $u^\beta \cap c_o$ is an FK space by Theorem 1.1.

<u>Remark</u>: There is no countable set $\{u_1, u_2, \cdots\}$ such that each $u_i \in \bigcap_{n=1}^{\infty} \ell^{1+\frac{1}{n}} - \ell$ and $\bigcup_{n=1}^{\infty} \ell^n = \bigcap_{i=1}^{\infty} u_i^\beta$, since $\bigcap_{i=1}^{\infty} u_i^\beta$ is an FK space.

<u>Application to Elementary Analysis.</u> Let $u^1, u^2, \cdots$ be sequences such that for each k, $\sum_{k=1}^{\infty} |u_k^i|^p < \infty$ for every $p > 1$, and $\sum_{k=1}^{\infty} |u_k^i| = \infty$.

There exists a sequence $x \in c_o$ such that $\sum_{k=1}^{\infty} |u_k^i x_k| < \infty$ for every i, and yet $\sum_{k=1}^{\infty} |x_k|^p = 0$ for every $p > 1$.

<u>Example</u> 1.2. $\bigcup_{n=1}^{\infty} \ell^n$ with the inductive limit topology by the inclusion maps is not metrizable.

<u>Proof</u>: Let $Y = \bigcap_{n=1}^{\infty} \ell^{1 + \frac{1}{n}}$. Y is weakly boundedly complete, for let $B \subset Y$ be bounded and weakly closed. Let x^δ be a Cauchy net in B with the weak topology. Then x^δ is a Cauchy net in s, so $x^\delta \to x$ in s. B is bounded in $\ell^{1 + \frac{1}{n}}$ for each n, say $\|b\|_{1+\frac{1}{n}} \leq M_n$ for every $b \in B$. Let m be a fixed integer.

$$\sum_{k=1}^{m} |x_k^\delta|^{1 + \frac{1}{n}} \to \sum_{k=1}^{m} |x_k|^{1 + \frac{1}{n}}.$$

$\sum_{k=1}^{m} |x_k^\delta|^{1+\frac{1}{n}} < [\|x^\delta\|_{1+\frac{1}{n}}]^{1+\frac{1}{n}} \leq (M_n)^{1+\frac{1}{n}}$. Since m is arbitrary, $x \in \ell^{1+\frac{1}{n}}$. Now for each integer n, let $M_n' = \max\left\{M_n, \|x\|_{1+\frac{1}{n}}\right\}$.

To see that Y is weakly boundedly complete, it remains to show that $x^\delta \to x$ weakly. Let $z \in Y' = \bigcup_{n=1}^{\infty} \ell^n = X$ say $z \in \ell^p$ and let $\epsilon > 0$. $|\sum_{i=1}^{\infty} z_i(x_i^\delta - x_i)| \leq R + R'$

where $R = |\sum_{i=1}^{m-1} z_i(x_i^\delta - x_i)|$ and

$$R' = |\sum_{i=m}^{\infty} z_i(x_i^\delta - x_i)|.$$

Now $$R' \leq \left\{\sum_{i=m}^{\infty} |z_i|^p\right\}^{1/p} \left\{\sum_{i=m}^{\infty} |x_i^\delta - x_i|^q\right\}^{1/q}$$

$$\leq 2M_{p-1}\,\epsilon \text{ for sufficiently large } m.$$

For fixed m, $R < \epsilon$ for all δ greater than or equal to some fixed δ_0.

Therefore, $x^\delta \to x \epsilon B$ and Y is weakly boundedly complete.

By [5], Theorem 20.1, Y is semi-reflexive. Thus, Y' with the strong topology is equal to Y. So X with the strong topology is equal to Y.

X is bornological by Corollary 2 of Theorem 4.1 of chapter two, so X is relatively strong. Then $\mathcal{J}_i$ is equal to the strong topology. The conclusion follows since the strong topology on the dual of a non-normable Frechet space is not metrizable ([5], Theorem 22.8; (13] § 13.5, # 11; [9] , page 21).

§ 2 The Lattice of BH Spaces.

Definition. A BH space is a Banach space continuously embedded in H.

The collection of BH spaces forms a lattice by Theorems 7.2 and 8.1 of chapter two.

Theorem 2.1. *Let $\{X_n\}$ be a sequence of BH spaces. $\bigcup_{n=1}^{\infty} X_n$ is a BH space if and only if there is a maximum X_n.*

Proof: Sufficieny is trivial.

Necessity follows from the Baire Category Theorem.

Corollary: Let $\{X_n\}$ be an increasing sequence of BH spaces. $\bigcup_{n=1}^{\infty} X_n$ is a BH space if and only if $\{X_n\}$ is stationary.

Theorem 2.2. *Let $\{X_n\}$ be a sequence of BH spaces.*

$I = \bigcap_{n=1}^{\infty} X_n$ <u>is a BH space if and only if there is an integer</u> m <u>such that</u> I <u>is closed in</u> $\bigcap_{n=1}^{m} X_n$.

<u>Proof</u>: Sufficiency is trivial.

Necessity - $N(0,1)$ in I is a neighborhood of 0 in $\bigvee_{n=1}^{\infty} \mathcal{J}_n$. Therefore, there exist $p_{i_1} \cdots p_{i_k}$, where p_{i_j} is the norm of X_{i_j}, and $\epsilon > 0$ such that $N(0,1) \supset \bigcap_{j=1}^{k} (p_{i_j} < \epsilon) \supset \bigcap_{i=1}^{m} (p_i < \epsilon)$. Therefore, there exists M such that $\|x\| \leq M \sum_{i=1}^{m} p_i(x)$ for every $x \in I$. Thus, the topology of I is weaker than that of $\bigcap_{i=1}^{m} X_i$. But since $I \subset \bigcap_{i=1}^{m} X_i$, the topology of I is stronger. Since I is complete, it is closed in $\bigcap_{i=1}^{m} X_i$.

<u>Corollary</u>: Let $\{X_n\}$ be a decreasing sequence of BH spaces. The following are pairwise equivalent:

1) $\bigcap_{n=1}^{\infty} X_n$ is a BH space

2) I is eventually closed in X_n

3) $\{cl_{X_n} I\}$ is stationary.

Proof: 1 implies 2 implies 3 is trivial.

3 implies 2 - Let $A_n = cl_{X_n} I$. Then $\{A_n\}$ is a decreasing sequence. For each n, $I \subset A_n \subset X_n$ so $\bigcap_{n=1}^{\infty} A_n = I$. There exists n_o such that for all $n \geq n_o$, $A_n = A_{n_o}$. So for all $n \geq n_o$, $I = \bigcap_{k=1}^{\infty} A_k = A_{n_o} = A_n$. Therefore, I is closed in X_n for all $n \geq n_o$.

2 implies 1 is trivial.

Theorem 2.3. *Let $\{X_n\}$ be a decreasing sequence of BH spaces. If I is a dense proper subset of each X_n, then $\{X_n\}$ has no inf in the lattice of BH spaces.*

Proof: Let $L = \inf\{X_n\}$. Let $x \in I$. $[x] = \text{span}\{x\} \subset X_n$ for each n. Also [x] is a BH space. Therefore, $[x] \subset L$, and $x \in L$. Hence for each n, $I \subset L \subset X_n$, so $I = L$. But I is not a BH space.

Example 2.1. $\{\ell^{1 + \frac{1}{n}}\}$ has no inf in the lattice of BK spaces.

§ 3 Normalized Order

Definition. Let X and Y be BH spaces. We write $X < Y$ when $X \subset Y$ and $\|x\|_X \geq \|x\|_Y$ for every $x \in X$.

Remark: If X and Y are BH spaces with $X \subset Y$, then X has an equivalent norm such that $X < Y$.

Proof: Since X has a finer topology than Y, there exists $\epsilon > 0$ such that $\|x\|_X \geq \epsilon \|x\|_Y$. The norm for X defined by $\frac{1}{\epsilon}\|\cdot\|_X$ satisfies the required property.

Example 3.1. Let X and Y be BH spaces. Then $X \cap Y < X < X + Y$ where $X \cap Y$ and $X + Y$ have their canonical norms.

Example 3.2. Let $(X, \|\cdot\|)$ be a Banach Space. Define X_n to be $(X, n\|\cdot\|)$. Then $X_{n+1} < X_n$ for each n. Also, the normalized infimum of $\{X_n\}$ is equal to $\{0\}$.

Proof: Clearly, $\{0\} < X_n$ for each n. Conversely, let Y be a BH space with $Y < X_n$ for each n. Let $y \in Y$. Then $\|y\|_Y \geq \|y\|_{X_n} = n \|y\|_X$ for every n. Thus, $\|y\|_X = 0$, and so $y = 0$.

Example 3.3. Let $X = H = m = \ell^\infty$. Let

$$\|x\|_n = \sum_{k=1}^{n} |x_k| + \sup_{i>n} |x_i|.$$

Each $\|\cdot\|_n$ is equivalent to $\|\cdot\|_o$, the natural norm for m. Define $X_n = (X, \|\cdot\|_n)$. Then $X_{n+1} < X_n$ for each n, and the normalized infimum of $\{X_n\}$ is equal to ℓ with its natural norm,

$$\|x\| = \sum_{k=1}^{\infty} |x_k|.$$

Proof: It is clear that $\ell < X_n$ for each n.

Let Y be a BH space with $Y < X_n$ for each n. Then $Y < \ell$, for let $y \in Y$. Since $\|y\|_Y \geq \|y\|_n$, we have $\|y\|_Y \geq \sum_{k=1}^{n} |y_k| + \sup_{k>n} |y_k| \geq \sum_{k=1}^{n} |y_k|$. Since this relationship holds for every n, $\sum_{k=1}^{\infty} |y_k| \leq \|y\|_Y$. Thus, $y \in \ell$, $\|y\|_\ell \leq \|y\|_Y$, and hence $Y < \ell$.

Example 3.4. In this example, we consider the ℓ^p spaces for $p > 1$ with their natural norms. We first notice three relationships

1. $\bigcap_{p>1} \ell^p = \left\{x : \|x\|_p < \infty \text{ for all } p > 1\right\}$

2. $\ell = \left\{x : \sup_{p>1} \|x\|_p < \infty\right\}$

3. $\|x\|_1 = \sup_{p>1} \|x\|_p$.

The first of these relationships is clear. It is also clear that the second follows from the third, so it is sufficient to prove 3.

Let $x \in \ell$. $\|x\|_p \leq \|x\|_1$ for every $p > 1$ by Jensen's Inequality.

Conversely, $\|x\|_p \geq \left\{\sum_{k=1}^{m} |x_k|^p\right\}^{1/p} \rightarrow \sum_{k=1}^{m} |x_k|$.

Therefore, $\sup_{p>1} \|x\|_p \geq \lim_{p\rightarrow 1} \|x\|_p \geq \sum_{k=1}^{m} |x_k|$. Since this is true for every integer m, $\sup \|x\|_p \geq \sum_{k=1}^{\infty} |x_k| = \|x\|_1$.

Finally, the normalized infimum of $\{\ell^p\}_{p>1}$ is equal to ℓ.

It follows from 3 that $\ell < \ell^p$ for all $p > 1$.

Conversely, let Y be a BK space with $Y < \ell^p$ for all $p > 1$. Let $y \in Y$. Then $\|y\|_Y \geq \|y\|_p$ for all $p > 1$, so $\|y\|_Y \geq \sup_{p>1} \|y\|_p = \|y\|_1$.

Thus, $y \in \ell$ and $\|y\|_Y \geq \|y\|_1$.

The next example shows that the normalized infimum of a sequence of BH spaces is not necessarily given by the supremum of the norms of those spaces.

Example 3.5. Let $H = R^{N^+}$. Let $X_n = H$ with the pseudo-norm $p_n(x) = |x_n|$. Then $\left\{x : \sup_n\{p_n(x)\} < \infty\right\}$ is equal to m, and the topology given by the supremum of the pseudonorms is the natural topology of m. However, the sup topology is the pointwise topology.

Theorem 3.1. (Aronszajn and Gagliardo, [1]).

A collection of BH spaces always has a normalized infimum.

Proof: Let $(X_\alpha, \|\cdot\|_\alpha)$ be a collection of BH spaces. Let $I = \bigcap_\alpha X_\alpha$, $Y = \left\{x \in I : \sup_\alpha \|x\|_\alpha < \infty\right\}$. Then Y with the norm $\|y\| = \sup \|y\|_\alpha$ is the normalized infimum of $\left\{X_\alpha\right\}$.

First, Y is complete, for let $\left\{x^n\right\}$ be a Cauchy sequence in Y. Then $\left\{x^n\right\}$ is Cauchy in each X_α, hence convergent in X_α, say $x^n \to x_\alpha$ in X_α. Then $x^n \to x_\alpha$ in H, so all the x_α's are equal. Let $x = x_\alpha$.

Now $x^n \to x$ in Y, for let $\epsilon > 0$ be given. There exists N such that $\|x^m - x^n\| < \epsilon$ for all $m,n \geq N$. Fix α. For all $m,n \geq N$ we have $\|x^m - x^n\|_\alpha < \epsilon$. Therefore, for all $m \geq N$ $\|x^m - x\|_\alpha \leq \epsilon$. Since this is true for all α we have $\sup_\alpha \|x^m - x\|_\alpha = \|x^m - x\| \leq \epsilon$ for all $m \geq N$. Thus, $x^m \to x$ in Y, and so Y is complete.

Finally, Y is the normalized infimum of $\left\{X_\alpha\right\}$. Clearly, $Y < X_\alpha$ for each α.

Conversely, let Z be a BH space with $Z < X_\alpha$ for all α. Let $z \in Z$. Then $\|z\|_Z \geq \|z\|_\alpha$ for each α, so $z \in Y$ and $\|z\|_Z \geq \|z\|_Y$.

Example 3.6. A countable collection of BH spaces need not have a normalized supremum even if $H = R$.

Let $X_n = (R, \frac{1}{n} |\cdot|)$. If Y is the normalized supremum of $\{X_n\}$, then for each $y \in Y$, $\|y\|_Y \leq \|y\|_n = \frac{1}{n} |y|$. Hence, $y = 0$ and so $Y = \{0\}$. But $X_n \subset Y$ for each n.

Theorem 3.2. Let X_α be a family of BH spaces with $X_\alpha < Z$ for some BH space Z and each α. Then $\{X_\alpha\}$ has a normalized supremum.

Proof: In any lattice in which every family has an infimum, every family bounded above has a supremum, for

$\sup\{X_\alpha\} = \inf\{Y : Y > X_\alpha \text{ for each } \alpha\}$.

Corollary: If H is a Banach space, the collection of all BH spaces X which satisfy $X < H$ is a complete lattice under the order $<$.

We next give a representation for the normalized supremum of a family of BH spaces bounded above.

With the same notation as in Theorem 3.2, the normalized supremum of $\{X_\alpha\}$ is equal to $N = \{z \in \cap\{Y : X_\alpha < Y \text{ for each } \alpha\}$ with $\|z\|_N = \sup\{\|z\|_Y : X_\alpha < Y \text{ for each } \alpha\} < \infty\}$.

Let $S = \left\{x \in Z : x = \sum_\alpha v_\alpha \text{ with } v_\alpha \in X_\alpha, \sum_\alpha \|v_\alpha\|_\alpha < \infty\right\}$.
For $x \in S$, let $\|x\| = \inf\left\{\sum_\alpha \|v_\alpha\|_\alpha : x = \sum_\alpha v_\alpha \text{ with } v_\alpha \in X_\alpha\right\}$.

First, $S \subset N$. Let $x \in S$. Then $x = \sum_\alpha v_\alpha$ with $v_\alpha \in X_\alpha$ and $\sum_\alpha \|v_\alpha\|_\alpha < \infty$. Let Y be a BH space with $X_\alpha < Y$ for each α. $\sum_\alpha \|v_\alpha\|_Y \leq \sum_\alpha \|v_\alpha\|_\alpha < \infty$. Since Y is complete, $\sum_\alpha v_\alpha$ is convergent in Y. Since H is Hausdorff, $\sum_\alpha v_\alpha$ must converge to x in Y. Thus, $x \in Y$. Also, $\|x\|_Y \leq \sum_\alpha \|v_\alpha\|_Y$, so $\|x\| \geq \|x\|_Y$. Since Y is arbitrary, $x \in N$ and $\|x\| \geq \|x\|_N$.

It is clear that for $x \in X_\alpha$, $\|x\| \leq \|x\|_\alpha$ since $x = \sum_\beta w_\beta$ where $w_\alpha = x$, $w_\beta = 0$ for $\beta \neq \alpha$. From these inequalities it will follow that for each α, $X_\alpha < S < N$ when we show that S is complete.

Let $\{x^n\}$ be a Cauchy sequence in S. We may assume that $x^o = 0$ and that $\sum_{n=1}^{\infty} \|x^n - x^{n-1}\| < \infty$. Let $x^n - x^{n-1} = \sum_\alpha v_\alpha^n$ with $\sum_\alpha \|v_\alpha^n\|_\alpha < \|x^n - x^{n-1}\| + 2^{\frac{1}{n}}$. Then

$$\sum_{n=1}^{\infty} \sum_{\alpha} \|v_\alpha^n\|_\alpha < \infty.$$

Since $v_\alpha^n = 0$ for all but countably many α, we re-index this double summation as

$$\sum_{n=1}^{\infty} \sum_{k=1}^{\infty} \|v_k^n\|_k < \infty.$$

Thus, $\sum_{n=1}^{\infty} \|v_k^n\|_k < \infty$ for each k.

Therefore, $\sum_{n=1}^{\infty} v_k^n$ is convergent in X_k, say $\sum_{n=1}^{\infty} v_k^n = w_k$.

$$\sum_{k=1}^{\infty} \|w_k\| \leq \sum_{k=1}^{\infty} \|w_k\|_k \leq \sum_{k=1}^{\infty} \sum_{n=1}^{\infty} \|v_k^n\|_k < \infty.$$

Therefore, $\sum_{k=1}^{\infty} w_k = y \in S$.

We now show that $x_n \to y$ in S.

$$\|x^m - y\| = \| \sum_{n=1}^{m} (x^n - x^{n-1}) - y\|$$

$$= \| \sum_{n=1}^{m} \sum_{k=1}^{\infty} v_k^n - \sum_{k=1}^{\infty} \sum_{n=1}^{\infty} v_k^n\|$$

$$= \| \sum_{k=1}^{\infty} \sum_{n=m+1}^{\infty} v_k^n\| \leq \sum_{k=1}^{\infty} \sum_{n=m+1}^{\infty} \|v_k^n\|$$

$$\leq \sum_{k=1}^{\infty} \sum_{n=m+1}^{\infty} \|v_k^n\|_k \to 0 \text{ as } m \to \infty.$$

Thus S is complete and since $X_\alpha < S < N$ for each α, $S = N$.

Chapter Five

§ 1 Garling's Completeness Theorem

Definition. An extended pseudonorm p on a linear topological space X is called lower semicontinuous if $(p \leq 1)$ is closed in X.

Theorem 1.1. (D. J. H. Garling, unpublished)

Let $(X,\sigma(\Phi))$ be a complete locally convex linear topological space. Let Ψ be a family of lower semicontinuous extended pseudonorms on $(X,\sigma(\Phi))$. Then $Y = \left\{x : p(x) < \infty \text{ for every } p \in \Psi\right\}$ is $\sigma(\Phi \cup \Psi)$ complete.

Proof: Let x^δ be a Cauchy net in $\sigma(\Phi \cup \Psi)$. Then x^δ is a Cauchy net in $\sigma(\Phi)$, hence is convergent. Say $x^\delta \to x$ in $\sigma(\Phi)$.

$x \in Y$, for let $p \in \Psi$. $p(x^\delta)$ is eventually bounded, say $p(x^\delta) \leq M$ for all $\delta \geq \delta_o$. Since $(p \leq M)$ is closed and $x^\delta \in (p \leq M)$ for all $\delta \geq \delta_o$, $p(x) \leq M$.

Finally, $x^\delta \to x$ in $\sigma(\Phi \cup \Psi)$. It is clearly sufficient to show that $p(x^\delta - x) \to 0$ for every $p \in \Psi$. Also we may assume without loss of generality that $x = 0$. Let $\epsilon > 0$ be given. There exists δ such that $p(x^\alpha - x^\beta) < \epsilon$ for all $\alpha \geq \delta$, $\beta \geq \delta$. Thus for $\alpha \geq \delta$, we have $p(x^\alpha - x) \leq \epsilon$.

Example 1.1. Let $X = s$, and suppose Ψ has one member.

Example 1.2. Let $\mathfrak{I}$ and $\mathfrak{I}'$ be locally convex topologies for a linear space X such that $\mathfrak{I}$ is complete and weaker than $\mathfrak{I}'$. If $\mathfrak{I}'$ has a local base of $\mathfrak{I}$ closed sets, then $\mathfrak{I}'$ is also complete.

Proof: Let Φ be the set of Gauges of $\mathfrak{I}$ closed, convex, balanced $\mathfrak{I}$ neighborhoods of 0. Let Ψ be the set of Gauges of $\mathfrak{I}$ closed, convex, balanced $\mathfrak{I}'$ neighborhoods of 0. Then $\mathfrak{I} = \sigma(\Phi)$, and $\mathfrak{I}' = \sigma(\Psi)$ since $\Phi \subset \Psi$.

§ 2 Complemented Subspaces

Let X and Y be disjoint FH spaces. Then the paranorm of $X + Y$ is given by $!x + y! = !x! + !y!$. Thus, $X + Y$ is an FH space with a paranorm which agrees on X and Y with their respective paranorms. A consequence is the following:

Theorem 2.1. *Let Z be an FH space, X and Y complementary subspaces of Z. If X and Y have any FH topologies, they must be closed in Z.*

Example 2.1. Let S be a subspace of c_o. If $c_o = \ell \oplus S$, then S has no FK topology.

Proof: ℓ is dense in c_o.

Example 2.2. Let S be a subspace of m with $m = c \oplus S$. Then S has no FK topology.

Proof: If S has an FK topology, it must be closed in m. This contradicts a theorem due to Phillips (see R.J. Whitley, American Mathematical Monthly, 73 (1966) pages 285-288).

Corollary 1. If X and Y are FH spaces with $X \subset Y$ and $\dim Y/X < \infty$, then X must be closed in Y.

Proof: $Y = X \oplus F$ where F is a finite dimensional subspace of Y.

A seemingly paradoxical consequence of this corollary is that if X and Y are FH spaces with X dense in Y, $\dim Y/X = \infty$. For example, $\dim c_0/\ell = \infty$.

Corollary 2. Let X be an FH space, and let $f \in X^{\#} - X'$. Then $f^{\perp}$ has no FH topology.

Corollary 3. Let X be an FH space which is not closed in H. Then there does not exist a maximal FH space including X as a closed subspace.

§ 3 Completion

Definition. An NH space is a non-complete normed subspace of H whose norm topology is stronger than that induced by H.

Definition. Let X be a linear topological space continuously embedded in H. We say Y is a BH completion of X if Y is a BH space including X and inducing on X the topology of X.

The question of whether every NH space has a BH completion is answered negatively by the following example.

Example 3.1. *Let* $(X, \mathcal{J})$ *be a BH space of infinite dimension. Let* $\|\cdot\|$ *be a norm giving* X *a topology strictly stronger than* $\mathcal{J}$. $(X, \|\cdot\|)$ *is an NH space. If* Y *is a BH space including* X *with the topology induced by* Y *on* X *equal to the topology given by* $\|\cdot\|$, *we have* $X \subset Y$ *and the topology induced by* Y *on* X *strictly stronger than* $\mathcal{J}$. *This is impossible by the remark following the definition of FH space on page 62.*

Remark: In order for an NH space X to have a completion it is necessary that X have no strictly stronger BH topology.

Definition. Let X be a linear topological space continuously embedded in H. We say Y is an FH completion of X if Y is an FH space including X and inducing on X the topology of X.

Theorem 3.1. *Let* X *be a linear metric space continuously embedded in* H. *Then* X *has an FH completion if and only if for every Cauchy sequence* $\{x_n\}$ *in* X *with* $x_n \to 0$ *in* H *we have* $x_n \to 0$ *in* X.

<u>Proof</u>: Necessity - Let $\bar{X}$ be an FH completion of X Let $\{x_n\}$ be a Cauchy sequence in X with $x_n \to 0$ in H. Then $x_n \to x$ in $\bar{X}$. Therefore, $x_n \to x$ in H, and so $x = 0$. So $x_n \to 0$ in $\bar{X}$, hence in X.

The proof of sufficiency is computational.

<u>Example</u> 3.2. Let H be a Frechet space, and let X be any infinite dimensional linear subspace. Then X can be given a linear metric topology embedding it continuously in H, and such that X has no FH completion.

<u>Proof</u>: Let p be a total paranorm for X which embeds X continuously into H. Let $f \in X^{\#} - (X,p)'$. Let $q(x) = p(x) + |f(x)|$. Then (X,q) has no FH completion, for we know that $(f = 1)$ is p dense in X. Therefore, there exists a sequence $\{x_n\}$ with $f(x_n) = 1$, $p(x_n) \to 0$. $\{x_n\}$ is q-Cauchy $[q(x^m - x^n) = p(x^m - x^n) + |f(x^m - x^n)| = p(x^m - x^n)]$, $x_n \to 0$ in H, but $q(x_n) \to 1$.

<u>Example</u> 3.3. E^{∞} can be given a linear metric topology with continuous coordinates and with no FK completion.

<u>Theorem</u> 3.2. <u>Let</u> X <u>be a linear metric sequence space with continuous coordinates, and let</u> $\bar{X}$ <u>be its completion. Then</u> X <u>has an</u> FK <u>completion if and only if</u> $\{\bar{P}_n\}$ <u>is total over</u> $\bar{X}$ <u>where</u> $\bar{P}_n$ <u>is the extension of the projection</u> P_n <u>to</u> $\bar{X}$.

<u>Proof</u>: Necessity - Trivial

Sufficiency - Let $\{x^n\}$ be Cauchy in X, $x^n \to 0$ in s. Then $x_n \to x$ in $\bar{X}$. $\bar{P}_j(x) = \lim \bar{P}_j(x^n) = \lim_{n\to\infty} x_j^n = 0.$

Since j is arbitrary, $x = 0$ and $x^n \to 0$ in X.

Chapter Five

Mixed Topologies and Two Norm Spaces

Definition. A mixed space (X,u,n) is a linear space X with a locally convex linear separated topology u and a norm n such that i. u is weaker than the topology of n

ii. n is u lower semi-continuous ($(n \leq 1)$ is u closed).

In case u is a paranorm, (X,u,n) is called a two norm space, and condition i says that u is n-continuous.

Example 1. Let X be a BH space such that the unit disc is H relatively closed in X. Let (Y,p) be any FH space including X. Then (X,p,n) is a two norm space.

Example 2. c_o, c, ℓ, m, ℓ^p all have the property of Example 1.

Proof: In m the unit disc is equal to $\bigcap_{n=1}^{\infty} \left\{x : |x_n| \leq 1\right\}$, hence is s relatively closed in m.

In ℓ^p, the unit disc is equal to $\left\{x : \sum_{k=1}^{\infty} |x_k|^p \leq 1\right\}$

$= \bigcap_{n=1}^{\infty} \left\{x : \sum_{k=1}^{n} |x_k|^p \leq 1\right\}$.

Remark 1. If (X,u) is complete, so is (X,n).

<u>Proof</u>: The topology of n is stronger than u and has a base of u-closed sets.

<u>Remark</u> 2. If u is a complete paranorm, then u and n are equivalent.

<u>Definition</u>. For any normed space X and a linear subspace $S \subset X'$, we define $\|\cdot\|_S$ on X to be $\|x\|_S = \|\hat{x}/_S\|$. Clearly, $\|\cdot\|_S \leq \|\cdot\|$.

We call S norming if $\|\cdot\|_S$ is equivalent to $\|\cdot\|$. See [13],§7.3.

Let (X,u,n) be a mixed space. Let $U = (X,u)'$ and let $N = (X,n)'$. Then $U \subset N$ by condition i, and moreover we have the following

<u>Theorem</u> 1. <u>U is a norming subspace of</u> N.

<u>Proof</u>: Let $x \in X$ with $n(x) = 1$. Let $0 < r < 1$. $(n \leq r)$ is u-closed, convex, and balanced, so then exists $f \in U : |f| \leq 1$ on $(n \leq r)$ and $f(x) = 1$. Thus, $\|f\| \leq \frac{1}{r}$, and $\|rf\| \leq 1$. $|rf(x)| = r = \hat{x}(rf)$. Therefore, $\|x/U\| \geq r$, and so $\|x\|_U \geq \|x\|$.

<u>Lemma</u> 1. Let X be a norm space, and let $f,g \in X'$ with $f(t) = g(t) = 1$. Then $\|f-g\| \leq a \, \|g/f^{\perp}\|$ where $a = 1 + \|f\| \, \|t\|$.

<u>Proof</u>: Computational.

Let (X,u,n) be a mixed space. Let $\{x^k\}$ be a sequence in X. We write $u(x^k) \to 0$ when $x^k \to 0$ in (X,u).

Definition. Let (X,u,n) be a mixed space, and let $\{x^k\}$ be a sequence in X. We say $x^k \to 0$ if $u(x^k) \to 0$ and $n(x^k)$ is bounded.

We say $x^k \to x \in X$ if $x^k - x \to 0$.

Facts.

i. If $x^k \to x$, $\lambda_k \to \lambda$, then $\lambda_k x^k \to \lambda x$.

ii. If $x^k \to 0$, and λ_k is a bounded sequence of scalars, then $\lambda_k x^k \to 0$.

iii. If $x^k \to x$, then for any subsequence x^{k_n}, $x^{k_n} \to x$.

iv. The sequential closure of a linear subspace is a linear subspace.

v. Let $f \in X^{\#}$. $f^{\perp}$ is either sequentially closed or sequentially dense.

Definition. Let $X^s = \{f \in X^{\#} : f$ is sequentially continuous, that is whenever $x^k \to 0$, $f(x^k) \to 0\}$.

It is clear that $U \subset X^s \subset N$.

Theorem 2. *Let $f \in X^{\#}$. If $f^{\perp}$ is sequentially closed, f is sequentially continuous.*

Proof: If f is not sequentially continuous, there exists $\{x^k\}$ with $x^k \to 0, \frac{1}{f(x^k)} < M$ (by fact iii). Choose t with

$f(t) = 1$. $t - \frac{x^k}{f(x^k)} \in f^\perp$. Since $x^k \to 0$, $t - \frac{x^k}{f(x^k)} \to t$.

So $t \in \overline{f^\perp} - f^\perp$.

<u>Lemma</u> 2. X^s is a closed subspace of N.

<u>Proof</u>: Let $f_k \in X^s$, $f_k \to f \in N$. Let $x_i \to 0$.

$$|f(x_i)| \leq |f(x_i) - f_k(x_i)| + |f_k(x_i)|.$$
$$\leq \|f - f_k\| \cdot M' + |f_k(x_i)|$$
$$< \epsilon + |f_k(x_i)| \text{ for sufficiently large } k.$$

But $|f_k(x_i)| \to 0$ as $i \to \infty$.

Therefore, $\lim |f(x_i)| \leq \epsilon$.

<u>Lemma</u> 3. Let (X,u,n) be a mixed space, and let A and B be sequentially closed subsets of X with $n[B]$ bounded. Then $A \cap B$ is u sequentially closed.

<u>Proof</u>: Let $x^k \in A \cap B$, $u(x^k - x) \to 0$. Each $x^k \in B$, so $\{x^k\}$ is n bounded. Thus, $x^k \to x$, and so $x \in A \cap B$.

<u>Theorem</u> 3. <u>Let</u> (X,u,n) <u>be a two norm space</u>. <u>Then</u> $X^s = cl_N U$.

<u>Proof</u>: $cl_N U \subset X^s$ by lemma 2.

Conversely, let $\epsilon > 0$. Let $f \in X^s$, and choose t with $f(t) = 1$. Let $C = f \cap (n \leq 1)$. C is u closed by lemma 3 and is convex and balanced. Therefore, there exists $g \in U$ with $|g| < \frac{\epsilon}{1+\|f\|n(t)}$ on C, $g(t) = 1$. By lemma 1, since $\|g/f^\perp\| < \frac{\epsilon}{1+\|f\|n(t)}$, $\|g-f\| < \epsilon$.

Theorem 4. Let X be a normed space, S a norm closed, norming subspace of X^*. Then a $\sigma(S)$ bounded set is norm bounded.

Proof: Let A be $\sigma(S)$ bounded. It is sufficient to show that every sequence in A is norm bounded. Let $\{x^k\} \subset A$. Define $\varphi : S \to m$ by $\varphi(f) = \{f(x^k)\}$. φ is continuous $[P_1 \circ \varphi (f) = f(x^1) = \hat{x}^1(f)$. So $P_1 \circ \varphi = \hat{x}1$ which is a continuous scalar function on S.]

Fix n. Choose $f \in S$ such that $\|f\| = 1$, $|f(x^n)| > k\|x^n\| - 1$ where k is chosen with $\|\cdot\|_s \geq k\|\cdot\|$. Then $k\|x^n\| \leq |f(x_n)| + 1 = [\varphi(f)]_n + 1 \leq \|\varphi\| + 1$. So $\{x^k\}$ is norm bounded.

Remark 1. The hypothesis that S be norm closed cannot be omitted. Let $X = c$, $S = E^\infty = \text{span}\{P_n\}$. S is norming since $\|x\|_S \geq |P_n(x)| = |x_n|$ for each n. Thus $\|x\|_S \geq \|x\|$.

Let $x^k = (\overbrace{0,0,\cdots 0}^{k},k,k,\cdots)$

$\|x^k\| = k$, but for every $f \in S$, $f(x^k) = 0$ eventually.

Remark 2. $\sigma(S)$ is not necessarily compatible with (X,X').

Remark 3. The converse of Theorem 4 holds even if S is not norm closed.

Proof: Suppose S is not norming. For every integer k, there exists x^k with $\|x^k\| > k$, $\|x^k\|_S < \frac{1}{k}$. $\{x^k\}$ is not norm bounded, but is $\sigma(S)$ bounded.

Remark 4. If $\|\cdot\|_S$ and $\|\cdot\|$ have the same bounded sets, S is norming since both norms give bornological topologies.

Remark 5. If S is closed in X*, $\sigma(S)$ and $\|\cdot\|_S$ have the same bounded sets.

Proof: Apply Theorem 4 to $(X, \|\cdot\|_S)$.

Theorem 5. (A. Wiweger)

Two norm convergence is topological, that is there exists a locally convex separated topology m such that $m(x^k) \to 0$ if and only if $x^k \to 0$.

Proof: Let $m = u \vee \sigma(X^s)$.

Let $x^k \to 0$. Then $u(x^k) \to 0$, and $f(x^k) \to 0$ for every $f \in X^s$, that is $m(x^k) \to 0$.

Conversely, let $m(x^k) \to 0$. Then $u(x^k) \to 0$ and $\{x^k\}$ is $\sigma(X^s)$ bounded. Hence, $\{x^k\}$ is norm bounded by Theorems 1, 3, and 4. Thus, $x^k \to 0$.

Remark. m is a Mazur topology.

Proof: Let f be sequentially continuous. Then $f \in X^s$, hence is $\sigma(X^s)$ continuous and thus is m continuous since m is stronger than $\sigma(X^s)$.

Chapter Six

§1 Coregular and Conull FK Spaces

Definition. An FK space X is called conservative if $c \subset X$. X is called coercive if $m \subset X$.

Definition. Let X be an FK space which includes $E^{\infty} \cup \{1\}$. We call X conull if $\sum_{k=1}^{n} \delta^{k} \to 1$ weakly in X. Otherwise, X is called coregular.

Example 1.1 s is conull. c is coregular since $1 \notin c_0 = \overline{E^{\infty}}$.

Remark 1. Let X be an FK space with $E^{\infty} \subset X$, $1 \notin X$. Then $X \oplus 1$ must be coregular.

Proof: X is a closed subspace of $X \oplus 1$ by

Remark 2. Let X be an FK space with $c_0 \subset X$. Then $\left\{\sum_{k=1}^{n} \delta^{k}\right\}$ is weakly Cauchy.

Proof: Let $f \in X'$. $f/c_0 \in c_0'$. Hence, $f(\sum_{k=1}^{n} \delta^{k}) = \sum_{k=1}^{n} f(\delta^{k})$ and so $\{f(\delta^{k})\} \in \ell$.

It follows that if X is conservative and weakly sequentially complete, X is conull. [$\sum_{k=1}^{n} \delta^{k}$ must converge, say $\sum_{k=1}^{n} \delta^{k} \to x$. But clearly $x_n = 1$ for each n.]

In particular, a conservative, coregular BK space cannot be reflexive.

§2 Generalization to FH Spaces.

Let H be a separated linear topological space such that H' is total over H. Let $\{w^n\}$ be a sequence in H, λ a point of H such that w^n converges weakly to λ.

Definition. A WH space is an FH space which includes λ and each w^n. We call a WH space X conull if w^n converges weakly to λ in X. Otherwise we call X coregular.

Example 2.1 $H = s$, $w^n = \sum_{k=1}^{n} \delta^k$, $\lambda = 1$.

Example 2.2 $H = L^2[0,1]$, $w^n = \frac{4}{\pi^2} \sum_{k=1}^{n} \frac{\sin(2k-1)x}{2k-1}$, $\lambda = 1$.

Let $C = \{f \in L^2:$ there exists a continuous function $g \in f\}$.

C is coregular, for define $\psi \in C'$ by $\psi(f) = g(0)$ where g is continuous, and $g \in f$. $\psi(\lambda) = 1$, while for each n, $\psi(w^n) = 0$.

Let Z be a fixed WH space such that $\{w^n\}$ is weakly Cauchy in Z, and such that w^n does not converge weakly to λ in Z.

Definition. An FH space is called conservative it it includes Z.

Theorem 2.1 *A conservative weakly sequentially complete FH space X is conull.*

Proof: In X, $\{w^n\}$ is weakly Cauchy, hence convergent. Say

$w^n \to \mu$ weakly in X. Then $w^n \to \mu$ weakly in H. Since $w^n \to \lambda$ weakly in H, and H' is total, $\lambda = \mu$.

<u>Example</u> 2.3 Let $X = \left\{x \in s : \left\{\frac{x_n}{n}\right\} \in \ell^2\right\}$, with $\|x\| = \|\left\{\frac{x_n}{n}\right\}\|_2$.

<u>Definition</u>. A WH space X is called permanent if λ is not in the linear closure of $\{w^n\}$.

X is called ρ-uniform if $w^n \to \lambda$ in X.

<u>Remark</u> 1. Permanent implies Coregular.

<u>Proof</u>: Let X be conull. Then $w^n \to \lambda$ weakly in X, so λ belongs to the weak linear closure of $\{w^n\}$ which is equal to the linear closure of $\{w^n\}$. Thus X is not permanent.

<u>Remark</u> 2. ρ-uniform implies conull.

<u>Example</u> 2.4 Coregular does not imply Permanent.

Let X be a Banach space with a Markucevic basis (biorthogonal system $\{b^n, f_n\}$ with $\{b^n\}$ fundamental, $\{f_n\}$ total) which is not a basis.

By [13], §12.4, Theorem 1 it is not a weak basis. Thus, there exists $x \in X$ such that $\sum_{n=1}^{k} f_n(x) b^n \not\to x$ weakly.

Let $t_k = f_k(x)$. We may assume each $t_k \neq 0$ [Otherwise consider $x + \sum_{k=1}^{\infty} u_k b^k$ where $u_k = 0$ if and only if $t_k \neq 0$, and $\|u_k b^k\| \leq \frac{1}{2^k}$].

Let $Y = \left\{\left\{\frac{f_n(z)}{t_n}\right\} : z \in X\right\}$

There is a natural mapping of X onto Y which carries an element $z \in X$ into $\left\{\frac{f_n(z)}{t_n}\right\}$. This mapping carries x into 1, and b^k into $\left\{\frac{\delta^k}{t_k}\right\}$.

In Y, 1 belongs to the linear closure of $\left\{\delta^k\right\}$, but $\sum_{k=1}^{n} \delta^k \not\to 1$ weakly.

<u>Example</u> 2.5 Conull does not imply ρ-uniform.

This is equivalent to the fact that $\sum_{k=1}^{n} \delta^k \to 1$ weakly does not imply that $\sum_{k=1}^{n} \delta^k \to 1$.

Let $X = \left\{x \in s : x_n - x_{n-1} \to 0\right\}$. X is congruent to c_o under the mapping which sends x into $\left\{x_n - x_{n-1}\right\}$. Under this mapping 1 is carried to δ^1, δ^k is carried into $\delta^k - \delta^{k+1}$.

In c_o, $\sum_{k=1}^{n} (\delta^k - \delta^{k+1}) \to \delta^1$ weakly since $\sum_{k=1}^{n} (\delta^k - \delta^{k+1}) = \delta^1 - \delta^{n+1} \not\to \delta^1$ since $\|\delta^{n+1}\| = 1$.

<u>Remark</u>. If X is an FH space such that $\left\{w^n\right\} \subset X$ and $\lambda \notin X$, $X \oplus \lambda$ is permanent.

<u>Proof</u>: X is closed in $X \oplus \lambda$ by Theorem 2.1 of Chapter 4.

<u>Example</u> 2.6 Let $Y = \left\{x \in s\text{: } \lim x_{2n} \text{ exists}\right\}$, $X = \left\{x \in s\text{: } \lim x_{2n} = 0\right\}$. Then $Y = X \oplus 1$. Y is permanent by the preceding remark.

Define $f(x) = \lim\left\{x_{2n}\right\}$. $f(\left\{\delta^k\right\}) = 0$, $f(1) = 1$.

Example 2.7 Let $X = \{x : x_n = O(n)\}$, and let $Y = \{x : x_n = o(n)\}$. X is congruent with m under the mapping which sends $x \in X$ into $\left\{\frac{x_n}{n}\right\}$. Y is congruent with c_o under the same mapping. Under this mapping 1 is carried to $\left\{\frac{1}{n}\right\}$, δ^k is carried to $\frac{\delta^k}{k}$. Since $\left\{\frac{1}{n}\right\} = \sum_{k=1}^{\infty} \frac{\delta^k}{k}$, $1 = \sum_{k=1}^{\infty} \delta^k$ in X and in Y.

In the following theorems, X and Y are FH spaces.

Theorem 2.2 *Let X be conull, $X \subset Y$. Then Y is conull.*

Proof: $w^n \to \lambda$ weakly in X, hence weakly in Y.

Theorem 2.3 *Let $X \subset Y$ with X closed in Y. Then X and Y are both coregular or both conull.*

Proof: X and Y have the same topology.

Corollary. If $\dim X/c < \infty$, X is coregular.

Proof: c is closed in X by Corollary 1 on page 79.

Remark 1. The union of a countable family of coregular FH spaces is coregular (if it is an FH space) because such a family must have a largest member.

Remark 2. The sum of two coregular FK spaces can be large. $X + Y = s$ is possible as is seen in the following example.

Example 2.8 Let $X = \{x \in s: \lim x_{2n} \text{ exists}\}$, $Y = \{x \in s: \lim x_{2n+1} \text{ exists}\}$

Theorem 2.4 The intersection of a countable family of conull FH spaces is conull.

Proof: Let $Y = \bigcap_{n=1}^{\infty} X_n$ with each X_n conull. Let $f \in Y'$. By [13], §4.4, #30, §11.3, #26 $f = \sum_{i=1}^{k} f_i$ where $f_i \in X_i'$. $f(w^n) = \sum_{i=1}^{k} f_i(w^n) \to \sum_{i=1}^{k} f_i(\lambda) = f(\lambda)$.

§3 Two Norm Spaces and Coregularity

Definition. Let X be a conservative FK space. For $f \in X'$, f/c is continuous on c. We define a linear functional ψ on X' by $\psi(f) = f(1) - \sum_{k=1}^{\infty} f(\delta^k)$.

Lemma. Let X be a conservative FK space. X is conull if and only if ψ vanishes on X'.

Proof: The proof follows immediately from the definition of conull.

It is well known that every $f \in c'$ has the form $f(x) = \psi(f) \lim x + \sum_{k=1}^{\infty} f(\delta^k)x_k$ (see [13], §6.4, Example 5). Hence, $|\psi(f)| = |\lim_{n\to\infty} [f(1) - \sum_{k=1}^{n} f(\delta^k)]| = |\lim_{n\to\infty} f(1- \sum_{k=1}^{n} \delta^k)| \leq \|f\|$, and so ψ is continuous on c^*. $\psi(\lim) = 1$, so $\|\psi\| = 1$. It is also true that $\psi \notin \hat{c}$.

Theorem 3.1 (Snyder's First Theorem) [7]

Let $(X, \|\,\|)$ be a conservative FK space. X is conull if and only if 1 belongs to the two norm sequential closure of E^∞.

<u>Proof</u>: Sufficiency - Let $x^n \to 1$, $x^n \in E^\infty$, with $|x_k^n| < M$ for each n,k. We wish to prove that for every $f \in X'$, $f(1) = \sum_{k=1}^{\infty} f(\delta^k)$. Consider $f(x^n) - \sum_{k=1}^{\infty} f(\delta^k)$

$$= f(\sum_{k=1}^{\infty} t_{n_k} \delta^k) - \sum_{k=1}^{\infty} f(\delta^k) = \sum_{k=1}^{\infty} (t_{n_k} - 1) f(\delta^k).$$

This series is uniformly convergent by the Weierstrass M test. Also, for each k, $x_k^n \to 1$ as $n \to \infty$.

Necessity - Suppose X is conull. It is sufficient to show that 1 belongs to the two norm sequential closure of c_o. [Say $y^n \in c_o$, $y^n \to 1$ in X, $|y_k^n| < M$ for each n,k. For each n choose $x_n \in E^\infty$ with $\|x_n - y_n\| < \frac{1}{n}$. Then Then a fortiori, $!x_n - y_n! \to 0$. So $x_n \to 1$ in X, and $\{x_n\}$ is bounded in c_o].

It is sufficient to show that c_o is not sequentially two norm closed in c [this implies that c_o is dense in c since it is a maximal subspace of c].

Let f be sequentially two norm continuous on c with $f = 0$ on c_o. f belongs to the two norm closure in c^* of X' by Theorem 3 of Chapter 5. Let $\epsilon > 0$ be given. There exists $g \in (c, !\,!)'$ with $\|f-g\| < \epsilon$. Extend g so that $g \in X'$. $|\psi(f)| = |\psi(f-g)| \leq \|\psi\| \, \|f-g\| < \epsilon \, \|\psi\|$. So $\psi(f) = 0$. But $\psi(f) = f(1)$, and hence $f = 0$ on c.

§4 Oscillation Spaces

Let $\{r_n\}$ be a strictly increasing sequence of positive integers with $r_1 = 1$. For $x \in s$, let
$0_n(x) = \max\{|x_u - x_v| : r_n \leq u \leq v \leq r_{n+1}\}$.

Definition. We define $\Omega(r) = \{x : 0_n(x) \to 0\}$.

Example 4.1 If $r_n = n$, $0_n(x)$ is just $|x_n - x_{n+1}|$.

For $x \in \Omega(r)$, we define $\|x\| = |x_1| + \sup |0_n(x)|$. Then $|x_n| \leq |x_n - x_{r_k}| + |x_{r_k} - x_{r_{k-1}}| + \cdots + |x_{r_2} - x_1| + |x_1|$ for $r_k \leq n \leq r_{k+1}$. Hence, $|x_n| \leq k\|x\|$ and so coordinates are continuous.

$\Omega(r)$ is complete, for let $\{x^n\}$ be a Cauchy sequence in $\Omega(r)$. Then for each k, $\{x_k^n\}$ is Cauchy. Let $x = \{\lim_{n\to\infty} x_k^n\}$. It is routine to check that $x \in \Omega(r)$, and that $x^n \to x$ in $\Omega(r)$.

$c \subset \Omega(r)$, for let $x \in c$, $\epsilon > 0$ be given. Choose a so that $|x_i - x_j| < \epsilon$ for all $i,j > a$. Choose b so that $r_n > a$ for all $n > b$. For $n > b$ we have $0_n(x) < \epsilon$.

Finally, $\Omega(r)$ is conull. To see this we show that 1 belongs to the two norm sequential closure of E^∞. This is equivalent to showing that there is a sequence $\{x^n\}$ in $\Omega(r)$ such that $x^n \to 0$ in $\Omega(r)$ and $x_k^n = 1$ eventually for each n. Let $x_k^n = \frac{i}{n}$ for $r_i < k \leq r_{i+1} \leq r_n$

$= 1$ for $r_n \leq k$.

<u>Lemma</u>. Let X be an FK space containing $1, E^\infty$, and a sequence $\{y^n\}$ in E^∞ such that $y^n \to 1$, $|y_k^n| < M$ for all n,k. Then there exists r such that $\Omega(r) \subset X + c_o$.

<u>Proof</u>: Let $x^n = 1 - y^n$, so that $x^n \to 0$ in X, $x_k^n = 1$ eventually for each fixed n, and $|x_k^n| < M$ for all n,k.

Let $r_1 = 1$. Inductively, assume that $r_1 \cdots r_{n-1}$, and $p_1 \cdots p_{n-1}$ have been defined. Choose $p_n > p_{n-1}$ so that $!\lambda x^{p_n}! < 2^{-n}$ for all $|\lambda| \leq 1$, and $|x_k^{p_n}| < 2^{-n}$ for $k = 1,2,\cdots r_{n-1}$. Choose $r_n > r_{n-1}$ so that $x_k^{p_r} = 1$ for all $k \geq r_n$.

Let $u \in \Omega(r)$. Let $v = \sum_{n=1}^\infty \lambda_n z^n$ where $z_n = x^{p_n}$, $\lambda_n = u_{r_n} - u_{r_{n-1}}$. $v \in X$ because $|\lambda_n| \leq 0_{n-1}(u) < 1$ eventually, so $!\lambda_n z^n! \leq 2^{-n}$. Let $w = u - v$. I assert that $w \in c_o$. Fix k. Choose m such that $r_m \leq k < r_{m+1}$.

$|w_k| = |u_k - v_k| \leq |u_k - \sum_{n=1}^m \lambda_n z_k^n| + |\lambda_{m+1} z_k^{m+1}| + \sum_{n=m+2}^\infty \lambda_n z_k^n|$.

$|u_k - \sum_{n=1}^m \lambda_n z_k^n| = |u_k - \sum_{n=1}^m \lambda_n|$ because $z_k^n = x_k^{p_n} = 1$ since $k \geq r_m \geq r_n$. But $|u_k - \sum_{n=1}^m \lambda_n| = |u_k - u_{r_m}| \leq 0_m(u) \to 0$ as $k \to \infty$.

$|\lambda_{m+1} z_k^{m+1}| \leq M|\lambda_{m+1}| \leq M\, 0_m(u) \to 0$ as $k \to \infty$.

$\sum_{n=m+2}^\infty |\lambda_n z_k^n| \leq \sum_{n=m+2}^\infty |\lambda_n| 2^{-n}$ because $|z_k^n| = |x_k^{p_n}| < 2^{-n}$ since $r_{n-1} \geq r_{m+1} \geq k$ ($n \geq m+2$, so $n-1 \geq m+1$).

But $\sum_{n=m+2}^\infty |\lambda_n| 2^{-n} \leq 2^{-m+1} \|\lambda\|_c \to 0$ as $k \to \infty$.

Corollary. (Snyder's Second Theorem)

A conservative conull FK space must include $\Omega(r)$ for some r.

Corollary. A conservative conull FK space must contain both bounded divirgent and unbounded sequences.

Remark 1. If $c \cap X$ is not closed in X, X must contain unbounded sequences ([13], §11.3, # 37).

For example, if X is conull and conservative, c is not closed in X hence X contains unbounded sequences.

Remark 2. X is called perfect if c is dense in X. A perfect space is either equal to c or contains unbounded sequences.

Remark 3. If c is not closed in X, X contains bounded divirgent sequences.

W. Meyer-Konig and K. Zeller. FK-Râume und Lűckenperfektheit, Mathematische Zeitschrift, 78 (1962), 143-148.

Remark 4. $\Omega(r) \cap m$ is always a non-separable subspace of m.

Corollary. If X is conull, $X \cap m$ is a non-separable subspace of m.

§ 5 Matrix Spaces

Let $A = (a_{nk})$ be a matrix.

Definitions. $d_A = \{x \in s : Ax \text{ exists}\}$. $c_A = \{x \in s : Ax \in c\}$. A is called conservative if c_A is conservative.

d_A is an FK space with $\{\delta^n\}$ as a basis ([13], §12.4, lemma 2, # 21). Thus d_A is ρ-uniform, hence conull.

Every $f \in c_A'$ has the representation
$f(x) = \alpha \lim_A x + \sum_{r=1}^{\infty} t_r(Ax)_r + \sum_{r=1}^{\infty} \beta_r x_r$ where $\lim_A x = \lim (Ax)$ and $\sum_{r=1}^{\infty} |t_r| < \infty$. ([13], § 12.4, Theorem 6).

If A is conservative $\sum_{r=1}^{\infty} |\beta_r| < \infty$, since $\sum_{r=1}^{\infty} \beta_r x_r$ converges for every $x \in c$.

It is well known that if A is conservative,
$\|A\| = \sup \sum_{k=1}^{\infty} |a_{nk}| < \infty$.

Remark. $E^{\infty} \subset c_A$ if and only if A has convergent columns. $1 \in c_A$ if and only if A has convergent row sums.

Definition. Let A be conservative. Recall that for $f \in c_A'$, $\psi(f) = f(1) - \sum_{k=1}^{\infty} f(\delta^k)$. We define

$$\chi(A) = \psi(\lim_A) = \lim_{n\to\infty} \sum_{k=1}^{\infty} a_{nk} - \sum_{k=1}^{\infty} \lim_{n\to\infty} a_{nk}.$$

Lemma. Let $f \in c_A'$. Then $\psi(f) = \alpha\, \chi(A)$ where α is given in the representation of f.

Proof: $\psi(f) = f(1) - \sum_{k=1}^{\infty} f(\delta^k) = \alpha \lim_A 1 + \sum_{r=1}^{\infty} t_r \sum_{k=1}^{\infty} a_{rk}$

$+ \sum_{r=1}^{\infty} \beta_r - \alpha \sum_{k=1}^{\infty} \lim_A(\delta^k) - \sum_{k=1}^{\infty} \sum_{r=1}^{\infty} t_r a_{rk} - \sum_{r=1}^{\infty} \beta_r = \alpha \chi(A)$ since

$\sum_{r=1}^{\infty} \sum_{k=1}^{\infty} |t_r a_{rk}| \leq \|A\| \sum_{r=1}^{\infty} |t_r| < \infty.$

Theorem 5.1 c_A *is coregular if and only if* $\chi(A) \neq 0$.

Proof: Sufficiency - $\lim_A 1 \neq \sum_{k=1}^{\infty} \lim_A \delta^k$, hence $\sum_{k=1}^{n} \delta^k$ does not converge weakly to 1.

Necessity - If $\chi(A) = 0$, $\psi(f) = \alpha \chi(A) = 0$ for every $f \in c_A'$. Hence c_A is conull by the lemma on page 94.

BIBLIOGRAPHY

1. N. Aronszajn and E. Gagliardo. "Interpolation spaces and interpolation methods." Annali di Mat. 68(1965) 51-118

2. N. Bourbaki. Espaces vectoriels topologiques. (Hermann, 1953).

3. R. E. Edwards. Functional Analysis. (Holt-Rinehart-Winston, 1965)

4. J. L. Kelley. General Topology. (Van Nostrand, 1955).

5. J. L. Kelley and I. Namioka. Linear topological spaces. (Van Nostrand 1963).

6. Victor Klee. "Exotic topologies for linear spaces". (J. Novak. Ed.) Prague Symposium 1961. General topology and its relations to modern analysis and algebra.

7. A. K. Snyder. "Conull and coregular FK spaces". Math. Zeitschrift 90(1965) 376-381.

8. Wendy Robertson. "Completions of topological vector spaces". Proc. Lond. Math. Soc. 8(1958) 242-257.

9. A. P. and W. Robertson. Topological vector spaces. (Cambridge 1964)

11. J. D. Weston. "The principle of equicontinuity for topological vector spaces." Proc. Durham Phil. Soc. 13 (1957) 1-5

12. J. D. Weston. "On the comparison of topologies". Journal Lond. Math. Soc. 32(1957) 342-354. "Incomplete subspaces of a Hausdorff space" Archiv der Math. 10(1959) 40-41.

13. A. Wilansky. Functional Analysis. (Blaisdell 1964).

16. V. Ptak. "Completeness and the open mapping theorem." Bulletin de la Societe Mathematique de France 86(1958) 41-74.

17. D. R. Anderson. "On connected irresolvable Hausdorff spaces". Proceedings of the American Mathematical Society 16(1965) 463-466.

18. J. Dugundji. Topology. (Allyn and Bacon 1966).

19. Victor Klee and S. Kakutani. "The finite topology of a linear space". Archiv der Mathematik 14(1963) 55-58.

20. G. Kothe. "Die Quotientraume eines linearen vollkommenen Raumes". Math. Zeitschrift 51(1947) 17-55.

Offsetdruck: Julius Beltz, Weinheim/Bergstr.

Lecture Notes in Mathematics

Bisher erschienen/Already published

Vol. 1: J. Wermer, Seminar über Funktionen-Algebren. IV, 30 Seiten 1964. DM 3,80

Vol. 2: A. Borel, Cohomologie des espaces localement compacts d'après J. Leray. IV, 93 pages. 1964 DM 9,–

Vol. 3: J. F. Adams, Stable Homotopy Theory. 2nd. revised edition. IV, 78 pages. 1966. DM 7,80

Vol. 4: M. Arkowitz and C. R. Curjel, Groups of Homotopy Classes. 2nd. revised edition. IV, 36 pages. 1967. DM 4,80

Vol. 5: J.-P. Serre, Cohomologie Galoisienne. Troisième édition. VIII, 214 pages. 1965. DM 18,–

Vol. 6: H. Hermes, Eine Termlogik mit Auswahloperator. IV, 42 Seiten. 1965. DM 5,80

Vol. 7: Ph. Tondeur, Introduction to Lie Groups and Transformation Groups. VIII, 176 pages. 1965. DM 13,50

Vol. 8: G. Fichera, Linear Elliptic Differential Systems and Eigenvalue Problems. IV, 176 pages. 1965. DM 13,50

Vol. 9: P. L. Ivănescu, Pseudo-Boolean Programming and Applications. IV, 50 pages. 1965. DM 4,80

Vol. 10: H. Lüneburg, Die Suzukigruppen und ihre Geometrien. VI, 111 Seiten. 1965 DM 8,–

Vol. 11: J.-P. Serre, Algèbre Locale. Multiplicités. Rédigé par P. Gabriel. Seconde édition. VIII, 192 pages. 1965. DM 12,–

Vol. 12: A. Dold, Halbexakte Homotopiefunktoren. II, 157 Seiten. 1966. DM 12,–

Vol. 13: E. Thomas, Seminar on Fiber Spaces. IV, 45 pages. 1966. DM 4,80

Vol. 14: H. Werner, Vorlesung über Approximationstheorie. IV, 184 Seiten und 12 Seiten Anhang. 1966. DM 14,–

Vol. 15: F. Oort, Commutative Group Schemes. VI, 133 pages. 1966. DM 9,80

Vol. 16: J. Pfanzagl and W. Pierlo, Compact Systems of Sets. IV, 48 pages. 1966. DM 5,80

Vol. 17: C. Müller, Spherical Harmonics. IV, 46 pages 1966. DM 5,–

Vol. 18: H.-B. Brinkmann und D. Puppe, Kategorien und Funktoren. XII, 107 Seiten. 1966 DM 8,–

Vol 19: G. Stolzenberg, Volumes, Limits and Extensions of Analytic Varieties. IV, 45 pages. 1966. DM 5,40

Vol. 20: R. Hartshorne, Residues and Duality. VIII, 423 pages 1966. DM 20,–

Vol. 21: Seminar on Complex Multiplication. By A. Borel, S. Chowla, C. S. Herz, K. Iwasawa, J.-P. Serre. IV, 102 pages. 1966. DM 8,–

Vol 22: H. Bauer, Harmonische Räume und ihre Potentialtheorie. IV, 175 Seiten. 1966 DM 14,–

Vol. 23: P. L. Ivănescu and S. Rudeanu, Pseudo-Boolean Methods for Bivalent Programming. 120 pages. 1966. DM 10,–

Vol. 24: J. Lambek, Completions of Categories. IV, 69 pages. 1966 DM 6,80

Vol. 25: R. Narasimhan, Introduction to the Theory of Analytic Spaces. IV, 143 pages. 1966. DM 10,–

Vol. 26: P.-A. Meyer, Processus de Markov. IV, 190 pages. 1967. DM 15,–

Vol. 27: H. P. Künzi und S.T. Tan, Lineare Optimierung großer Systeme. VI, 121 Seiten. 1966. DM 12,–

Vol. 28: P. E. Conner and E. E. Floyd, The Relation of Cobordism to K-Theories. VIII, 112 pages. 1966 DM 9,80

Vol 29: K. Chandrasekharan, Einführung in die Analytische Zahlentheorie. VI, 199 Seiten. 1966. DM 16,80

Vol. 30: A. Frölicher and W. Bucher, Calculus in Vector Spaces without Norm. X, 146 pages. 1966. DM 12,–

Vol. 31: Symposium on Probability Methods in Analysis. Chairman: D.A. Kappos. IV, 329 pages. 1967. DM 20,–

Vol. 32: M. André, Méthode Simpliciale en Algèbre Homologique et Algèbre Commutative. IV, 122 pages. 1967. DM 12,–

Vol. 33: G. I. Targonski, Seminar on Functional Operators and Equations. IV, 112 pages. 1967. DM 10,–

Vol. 34: G. E. Bredon, Equivariant Cohomology Theories. VI, 64 pages. 1967. DM 6,80

Vol. 35: N. P. Bhatia and G. P. Szegö, Dynamical Systems: Stability Theory and Applications. VI, 416 pages. 1967. DM 24,–

Vol. 36: A. Borel, Topics in the Homology Theory of Fibre Bundles. VI, 95 pages. 1967. DM 9,–

Bitte wenden / Continued

Vol. 37: R. B. Jensen, Modelle der Mengenlehre. X, 176 Seiten. 1967. DM 14,–

Vol. 38: R. Berger, R. Kiehl, E. Kunz und H.-J. Nastold, Differentialrechnung in der analytischen Geometrie. IV, 134 Seiten. 1967. DM 12,–

Vol. 39: Séminaire de Probabilités I. IV, 187 pages. 1967. DM 14,–

Vol. 40: J. Tits, Tabellen zu den einfachen Lie Gruppen und ihren Darstellungen. VI, 53 Seiten. 1967. DM 6,80

Vol. 41: R. Hartshorne, Local Cohomology. VI, 106 Seiten. 1967. DM 10,–

Vol. 42: J. F. Berglund and K. H. Hofmann, Compact Semitopological Semigroups and Weakly Almost Periodic Functions. VI, 146 Seiten. 1967. DM 12,–

Vol. 43: D. G. Quillen, Homotopical Algebra. VI, 157 Seiten. 1967. DM 14,–

Vol. 44: K. Urbanik, Lectures on Prediction Theory. II, 50 Seiten. 1967. DM 5,80